Mostafa Zahri

Energy-based model of condensing on metric spaces

Mostafa Zahri

Energy-based model of condensing on metric spaces

Modeling, analysis and simulations of condensing of particles

Südwestdeutscher Verlag für Hochschulschriften

Impressum/Imprint (nur für Deutschland/ only for Germany)
Bibliografische Information der Deutschen Nationalbibliothek: Die Deutsche Nationalbibliothek verzeichnet diese Publikation in der Deutschen Nationalbibliografie; detaillierte bibliografische Daten sind im Internet über http://dnb.d-nb.de abrufbar.
Alle in diesem Buch genannten Marken und Produktnamen unterliegen warenzeichen-, marken- oder patentrechtlichem Schutz bzw. sind Warenzeichen oder eingetragene Warenzeichen der jeweiligen Inhaber. Die Wiedergabe von Marken, Produktnamen, Gebrauchsnamen, Handelsnamen, Warenbezeichnungen u.s.w. in diesem Werk berechtigt auch ohne besondere Kennzeichnung nicht zu der Annahme, dass solche Namen im Sinne der Warenzeichen- und Markenschutzgesetzgebung als frei zu betrachten wären und daher von jedermann benutzt werden dürften.

Verlag: Südwestdeutscher Verlag für Hochschulschriften Aktiengesellschaft & Co. KG
Dudweiler Landstr. 99, 66123 Saarbrücken, Deutschland
Telefon +49 681 37 20 271-1, Telefax +49 681 37 20 271-0, Email: info@svh-verlag.de
Zugl.: Frankfurt, UNI, Diss., 2009

Herstellung in Deutschland:
Schaltungsdienst Lange o.H.G., Berlin
Books on Demand GmbH, Norderstedt
Reha GmbH, Saarbrücken
Amazon Distribution GmbH, Leipzig
ISBN: 978-3-8381-1081-3

Imprint (only for USA, GB)
Bibliographic information published by the Deutsche Nationalbibliothek: The Deutsche Nationalbibliothek lists this publication in the Deutsche Nationalbibliografie; detailed bibliographic data are available in the Internet at http://dnb.d-nb.de.
Any brand names and product names mentioned in this book are subject to trademark, brand or patent protection and are trademarks or registered trademarks of their respective holders. The use of brand names, product names, common names, trade names, product descriptions etc. even without a particular marking in this works is in no way to be construed to mean that such names may be regarded as unrestricted in respect of trademark and brand protection legislation and could thus be used by anyone.

Publisher:
Südwestdeutscher Verlag für Hochschulschriften Aktiengesellschaft & Co. KG
Dudweiler Landstr. 99, 66123 Saarbrücken, Germany
Phone +49 681 37 20 271-1, Fax +49 681 37 20 271-0, Email: info@svh-verlag.de

Copyright © 2009 by the author and Südwestdeutscher Verlag für Hochschulschriften Aktiengesellschaft & Co. KG and licensors
All rights reserved. Saarbrücken 2009

Printed in the U.S.A.
Printed in the U.K. by (see last page)
ISBN: 978-3-8381-1081-3

Preface

What is the importance of the energy as a rule of moves in systems of multi-agents? Which influences have the social and the spatial characteristics of agents' energy in a population? If we suppose that each agent wants to minimize his local energy produced from local interactions, what will happen after a large number of interactions? This work gives a new point of view concerning the rule of moves produced by local interactions in population. In several other contexts, scientists have modeled and analyzed the phenomena of multi-agents systems interactions in biological, ecological, and sociological systems. They have proved that the emergence behaviors of such populations have complex natural and social backgrounds. Our work is a contribution to the challenging of the modeling of population's emergence and dynamics. It presents an energy-based model of condensing sequences on metric spaces. The local epsilon energy of individuals presents a motif of local moves for all agents. Each singular move causes a reorganizing of all member of the group by decreasing the total energy, where the finial state is either condensing or segregation.

Frankfurt am Main 2009, M. Zahri

Acknowledgments

Grateful acknowledgments are given to Prof. Dr. Malte Sieveking, my chief supervisor, who gives me the opportunity to write this Dissertation in Frankfurt am Main, for the helpful discussions and the constructive remarks.

I wish to thank Prof. Dr. Dr. Ulrich Krause, who took time to read this Dissertation and to have been a member of my Ph.D jury.

Contents

Preface 1

Acknowledgments 3

List of Figures 6

List of Tables 9

Introduction 11

1 Condensing on finite metric space   **15**

    1.1 Condensing 15

    1.2 Convergence 18

    1.3 Random metric on finite metric space 22

        1.3.1 Extremal pseudometrics 22

        1.3.2 Algorithms for the construction of random metrics 26

        1.3.3 Numerical simulations of random metrics 31

    1.4 Numerical simulations of condensing sequences 36

        1.4.1 Simulations on an Euclidean finite metric space 37

        1.4.2 Simulations in a finite circular metric space 43

        1.4.3 Simulations with respect to a random metric 50

        1.4.4 Simultaneously condensing with respect to a random metric   58

2 Condensing in continuous metric space   **59**

    2.1 Condensing 59

2.2	Convergence		65
2.3	Numerical Simulations		67
	2.3.1	Condensing on the real line	68
	2.3.2	Condensing on the unit circle	73
	2.3.3	Condensing on the real plane	78
	2.3.4	Segregation sequences	84
	2.3.5	Simultaneously condensing on the unit circle	88

Concluding remarks 90

Bibliography 92

List of Figures

1.1 *Euclidean metric vector, NP=25, and 50.* 31

1.2 *Random metric vector (Uniform, Normal), NP=25, and 50.* 32

1.3 *Euclidean metric mesh, NP=25, and 50.* 33

1.4 *Random metric mesh (Uniform), NP=25.* 33

1.5 *Random metric mesh (Uniform), NP=50.* 34

1.6 *Random metric mesh (Normal), NP=25, and NP=50.* 35

1.7 *(Uniform) Random metric contour, NP=50.* 35

1.8 *(Normal) Random metric contour NP=50.* 36

1.9 *Condensing in an Euclidean finite metric space (sim. (a)).* 40

1.10 *Condensing in an Euclidean finite metric space (sim. (b))* 41

1.11 *Condensing in an Euclidean finite metric space (sim. (c)).* 42

1.12 *Condensing in a geodesic finite metric space (sim. (a))* 45

1.13 *Condensing in a geodesic finite metric space (sim. (b))* 46

1.14 *Condensing in a geodesic finite metric space (sim. (c))* 47

1.15 *Condensing in a geodesic finite metric space (sim. (a,b, c))* 49

1.16 *Condensing in a (random) metric space, for $\varepsilon = 0.182508$ (sim. (a))* 52

1.17 *Condensing in a (random) metric space, for $\varepsilon = 0.200761$ (sim. (b))* 53

1.18 *Condensing in a (random) metric space, for $\varepsilon = 0.254162$ (sim. (c))* 55

1.19 *Distances at limit for sim.(a) and (b)* 56

1.20 *Cardinal of support m^i at limit for sim.(a) and (b)* 56

LIST OF FIGURES

1.21 Energy of sim. (a), (b) and (c) in diff. (random) metric spaces. . . . 57

1.22 Simultaneously condensing with resp. 1000 RM, Mean energy (right). 58

2.1 Density of condensing measures on the real line (sim. (a)). 70

2.2 Density of condensing measures on the real line (sim. (b)) 71

2.3 Condensing of particles on a one dimensional manifold (sim. (a)). . 75

2.4 Condensing of particles on a one dimensional manifold (sim. (b)). . 76

2.5 Condensing of particles on a one dim. manifold (sim. (a) and (b)). . 77

2.6 Density of condensing of particles on real plane (sim. (a)). 80

2.7 Contour of the density of the condensing on real plane (sim. (a)). . 81

2.8 Density of condensing of particles on real plane (sim. (b)) 82

2.9 Contour of the density of the condensing on real plane (sim. (b)) . 83

2.10 Density (right) and spatial position (left) of segregation. (sim. (a)). 86

2.11 Density (right) and spatial position (left) of segregation. (sim. (b)). 87

2.12 Simultaneously condensing on the unit circle for the average model. 89

2.13 Simultaneously condensing on the unit circle for the energy model. . 89

List of Tables

1.1 Results of simulations in an (Euclidean) finite metric space. 38

1.2 Results of simulations in a finite metric space (subset of S^1). 44

1.3 Results of simulations (random) finite metric space. 51

2.1 Results of simulations in the real line. 69

2.2 Results of simulations in S^1. 73

2.3 Results of two simulations on the real plane. 78

2.4 Results of two simulations in $I\!R^2$ (segregation). 84

2.5 Results of simultaneously condensing sequences in S^1. 88

Introduction

The present study is motivated by work of Hegselmann and Krause (HK) on consensus dynamics [15], where agents simultaneously move to the barycenter of all agents in an epsilon neighborhood. The final state may be consensus, where all agents meet at the same position or grouping several classes of agents such that all agents in the same class maintain the same position and agents in different classes are at a distance greater than or equal to epsilon. In this work, we are interested to extend the HK model given as example by [16, 17] to a metric space. Observe that, the barycenter of a measure m minimizes the epsilon-energy of a position:

$$e_\varepsilon(x,m) = \int_{d(x,y)\leq\varepsilon} d^2(x,y) m(dy),$$

where for the HK dynamics, $d(\cdot,\cdot)$ is an Euclidean distance. This observation is the starting point for our present study to generalize the HK model. We

1. replace the Euclidean space by an arbitrary metric space, and

2. let the agents move to where the local energy is minimal within an epsilon neighborhood.

Because of the second claim, this does not generalize HK dynamics, as it is already demonstrated by two agents and Euclidean metric: Two agents may decrease the energy to zero by jumping either to the same place, or to different places if the distance exceeds epsilon. It is important to note that our dynamics, because of claim 2, is not a deterministic one. Furthermore, the convergence of the process of "condensing", as we call it, is not guaranteed. This fact can be seen in the case of two agents, they may exchange there position for ever, with periodic local energy. In order to be able to prove convergence, we deviate from HK in another way

3. Agents do not move simultaneously but one at a time in an arbitrary order. By doing so, they decrease the total epsilon energy:

$$E_\varepsilon(m) = \int_X e_\varepsilon(x,m) m(dx),$$

which guaranties the convergence and in fact zero energy after finitely many steps. It is also important to note that the arbitrary order of action of different agents introduces yet another source of indeterminacy. Such indeterminacy gives the opportunity for stochastic investigations, which however are not part of the present study. Our concern in this thesis is the introduction of a new class of dynamical systems-together with some elementary analysis and a number of numerical simulations. Several authors explain consensus dynamics in the context of emergence, see for example [9, 20]. We shall briefly discuss in how far our model responds to the challenge of emergence.

Ants live in large populations. These population show a complicated and strict division of labor for the individual ant, which on the one hand is not determined by the genetic structure of the single ant, and on the other hand makes the whole population react effectively to all kinds of events as if steered by some clever and experienced brain, which however does not exist. The division of labor, which makes an ant become a soldier, and another one a messenger is called emergent. It is very strictly and very stable, but one does not detect it as a program in the individual. Another example is demonstrated in the case of cells. Although all cells in the human being have the same DNA they diversify into different functions to build the human body. Therefore, the organization which results in this diversification is called emergent. That one cell becomes a brain cell and another one a liver cell may depend but on its ambient: for example the pressure from faster growing cells on top of one cell may cause it to become a brain cell finally. Thus, the fate of a single cell seems to be largely at random, whereas the the result: the human being is very well defined stable and obviously fixed in advance. How is this to be understood? This is the challenge of emergence - as I see it - and we will briefly discuss in how far our model models emergence. There is an emergent pattern: segregation into positions with two of them at distance grater than ε. However, the number of positions and the distribution of individuals onto there positions seem to be random.

In this study, we are interested

1. To extend the HK model presented in [16] to an arbitrary metric space.

2. To perform theoretical analysis for this model such as convergence theorems.

3. To present numerical results in finite and continuous metric spaces, with a special application in the condensing of particles. We also simulate the HK model with respect to a large number of random metrics and on the unit circle as a one dimensional manifold.

The simulation in a finite metric space or the so called n points metric space motivated us to introduce the notion of "random metrics". These are constructed as a positive linear combination of extremal pseudometrics. The theoretical and numerical constructions of class of such metric are obtained by using the results of cut or split metrics. For more details about split metrics, we refer to the studies by Dress and al. [2, 11, 12]. More generally, we propose an algorithm to construct any random metric in a finite metric space as a solution of linearly independent of a system of equations. This thesis is structured in two principal chapters. The first one presents the construction of condensing sequences in a finite metric space, where we also introduce some methods to construct random metrics. The second chapter extends the condensing model in a continuous metric space. The last section of this thesis contains general concluding remarks, outlooks and some open problems.

CHAPTER 1

Condensing on finite metric space

Local interactions between particles of a collection causes all particles to reorganize in new positions. The purpose of this chapter is to construct a model, which describes such a phenomena in finite metric space (FMS). In order to simulate the condensing sequences in FMS, we introduce the notion of random metric. Our model is analyzed and simulated in many finite metric spaces and also simulated with respect to a random metric.

1.1 Condensing

Let (X, d) be a finite metric space with metric d. A non negative measure m on X is represented by a function $m : X \to [0, \infty)$ in the obvious way. Denote by $M_+(X)$ the set of all positive measures on X. A measure $m \in M_+(X)$ is given as

$$m = \sum_{x \in X} m(x)\delta_x = \sum_{x \in S(m)} m(x)\delta_x,$$

where δ_x denotes the Kronecker symbol and by $S(m)$ we denote the support of m given as

$$S(m) := \{y \in X | m(y) > 0\}. \tag{1.1}$$

Definition 1.1.1. Fix a real number $\varepsilon > 0$. The ε-energy of m is

$$E_\varepsilon(m) = \sum_{d(x,y) \leq \varepsilon} m(x)m(y)d^2(x,y). \tag{1.2}$$

1.1. Condensing

The $\varepsilon-$ energy of point $a \in X$ with respect to m is

$$e_\varepsilon(a, m) = \sum_{d(a,y) \leq \varepsilon} m(y) d^2(a, y). \tag{1.3}$$

Let a pair $(a, a^*) \in X \times X$ operates on the set $M_+(X)$ of nonnegative measures on X as $m \to (a, a^*, m)$, where

$$(a, a^*, m)(x) = \begin{cases} m(x); & \text{if } x \notin \{a, a^*\}, \\ 0; & \text{if } x = a, \\ m(a) + m(a^*); & \text{if } x = a^*. \end{cases}$$

Definition 1.1.2. A pair of masses (m, m^*) is called an $\varepsilon-$move, if there is a pair $(a, a^*) \in X \times X$ such that:

(i) $m^* = (a, a^*, m)$,

(ii) $d(a, a^*) \leq \varepsilon$.

(iii) $e_\varepsilon(a, m) + m(a) d^2(a, a^*) > e_\varepsilon(a^*, m)$.

We are interested in sequences of non negative measures $m^1, m^2, \ldots$ satisfying the conditions above. Such a sequence $m^1, m^2, \ldots$ is called $\varepsilon-$condensing.

Remark 1.1.1. It is important to note that the condition (iii) of definition 1.1.2 is equivalent to

$$a \text{ moves to } a^* \text{ and } e_\varepsilon(a^*, m^*) < e_\varepsilon(a, m). \tag{1.4}$$

Clearly for every $a, a^* \in S(m)$ if $d(a, a^*) \leq \varepsilon$, then either (a, a^*, m) or (a^*, a, m) is an $\varepsilon-$move. Therefore, whenever $E_\varepsilon(m) > 0$ there is an $\varepsilon-$move (m, m^*). Thus, for every finite m with non vanishing energy, there is an $\varepsilon-$condensing sequence $m^1, m^2, \ldots$. Our theorem says that such a sequence is finite.

More generally is the following definition:

1.1. Condensing

Remark 1.1.2. Let (X, d) be a finite metric space. A mapping $f : S(m) \to X$ is said to be condensing if for every $y \in X$ either $f(y) = y$ or

$$\sum_{d(z, f(y)) \leq \varepsilon} m(z) d^2(z, f(y)) < m(y) d^2(y, f(y)) + \sum_{d(z,y) \leq \varepsilon} m(z) d^2(y, z). \quad (1.5)$$

$m^1, m^2, \ldots$ is said to be condensing if for every i there is a mapping f_i which is condensing and such that

$$m^{i+1} = f_i(m^i). \quad (1.6)$$

A condensing mapping f is said to be singularly condensing if $f(y) \neq y$ for only one $y \in S(m)$. A condensing sequence is called singularly if every f_i is singularly. f is said to be simultaneously condensing, if the condition (1.5) is required for more than one elements of $S(m)$. Note that for the image measure of a singularly condensing mapping as, we find

$$f : M_+(X) \longrightarrow M_+(X); \quad (1.7)$$

$$f(m)(x) := \begin{cases} m(x), & \text{if } f(x) \notin S(m), \\ m(x) + m(z), & \text{if } f(x) = z \in S(m). \end{cases}$$

Remark 1.1.3. The simultaneously displacement sequence are studied in another context in the literature by synchronous communication, moves and reactions, for example, we refer to the models studied in [17, 19, 21]. Therefore, the condensing mapping given by (1.5) is a generalization of all types of reactions, namely singular and simultaneous.

It is also important to note that the resulting measure of a condensing sequence depends not only on the initial measure, but also on the order of succession of the particles reactions. Hence, we introduce a random range of the order of reaction of particles. The same idea was proposed and developed by Sieveking [21] in the case of the real line.

1.2 Convergence

The purpose of this section is to prove the following theorem about convergence of ε-condensing sequences:

Theorem 1.2.1. A singularly ε-condensing sequence is finite.

The proof will be a consequence of the following lemmas.

Lemma 1.2.1. Let $m \in M_+(X), a, a^* \in X$ such that $d(a, a^*) \leq \varepsilon$. Then

$$E_\varepsilon(m) - E_\varepsilon(m^*) = 2m(a)\big[e_\varepsilon(a, m) - e_\varepsilon(a^*, m) + m(a)d^2(a, a^*)\big], \quad (1.8)$$

where $m^* = (a, a^*, m)$.

Proof. To simplify, we use the following notation

$$I_m := \sum_{d(x,y)\leq\varepsilon;\{x,y\}\cap\{a,a^*\}=\emptyset} m(x)m(y)d(x,y)^2 \quad (1.9)$$

Let us compute the energy of m:

$$\begin{aligned} E_\varepsilon(m) &= \sum_{d(x,y)\leq\varepsilon} m(x)m(y)d^2(x,y) \\ &= I_m + 2m(a)\sum_{d(a,x)\leq\varepsilon} m(y)d^2(x,y) + 2m(a^*)\sum_{d(a^*,y)\leq\varepsilon} m(y)d^2(a^*,y) \\ &\quad -2m(a)m(a^*)d^2(a,a^*) \end{aligned}$$

then

$$\begin{aligned} E_\varepsilon(m) &= I_m + 2m(a)e_\varepsilon(a,m) + 2m(a^*)e_\varepsilon(a^*,m) \\ &\quad -2m(a)m(a^*)d^2(a,a^*). \end{aligned} \quad (1.10)$$

Similarly for $m^* = (a, a^*, m)$:

$$E_\varepsilon(m^*) = I_{m^*} + 2m^*(a)e_\varepsilon(a, m^*) + 2m^*(a^*)e_\varepsilon(a^*, m^*)2m^*(a)m^*(a^*)d^2(a,a^*)$$

1.2. Convergence

Note that
$$I_m = I_{m^*}; \quad m(a) = 0; \quad m^*(a^*) = m(a) + m(a^*).$$

and
$$e(a^*, m^*) = e(a^*, m) - m(a)d^2(a^*, a). \tag{1.11}$$

Therefore
$$E_\varepsilon(m^*) = I_m + 2(m(a) + m(a^*))e_\varepsilon(a^*, m^*) \tag{1.12}$$
$$= I_m + 2(m(a) + m(a^*))(e(a^*, m) - m(a)d^2(a^*, a)).$$

and from (1.10) and (1.12) it follows:
$$E_\varepsilon(m) - E_\varepsilon(m^*) = 2m(a)\big[e_\varepsilon(a, m) - e_\varepsilon(a^*, m) + m(a)d^2(a, a^*)\big]. \tag{1.13}$$

□

Lemma 1.2.2. For $m \in M_+(X)$ let $n(m)$ be the number of elements $a \in X$ such that $m(a) > 0$. If $m^1, m^2, \ldots$ is a sequence of measures on X which is singularly and ε-condensing, then

1. $i \to E_\varepsilon(m^i)$ is strictly decreasing

2. $i \to n(m^i)$ is decreasing

Proof. The first claim follows from lemma 1.2.1. To show the second let $S(m) = \{x \in X | m(x) > 0\}$ be the support of the measure m. Consider
$$m^* = (a, a^*, m).$$
If $a \notin S(m)$ then $S(m) = S(m^*)$ and $n(m) = n(m^*)$. If $a \in S(m)$ and $a^* \in S(m)$ then $S(m^*) = S(m) \setminus \{a\}$ and $n(m^*) < n(m)$. If $a \in S(m)$, $a^* \notin S(m)$ then
$$S(m^*) = (S(m) \setminus \{a\}) \cup \{a^*\},$$
and again $n(m) = n(m^*)$.

1.2. Convergence

□

Proof. of theorem 1.2.1. Let $m^1, m^2, \ldots$ be an infinite sequence of measures, which is ε-condensing. Because of the preceding lemma, we may assume that

$$i \to n(m^i)$$

is constant. Hence, for every i The measure m^{i+1} is a permutation of m^i i.e.

$$m^{i+1} = m^i \circ \pi_i, \tag{1.14}$$

where $\pi_i : X \to X$ is a permutation of X. Therefore,

$$m^i = m^1 \circ \pi_1 \circ \ldots \pi_{i-1}. \tag{1.15}$$

As the group of permutations of X is finite, there exist a natural numbers $i, k > 0$ such that

$$\pi^1 \circ \ldots \pi^i = \pi^1 \circ \ldots \pi^{i+k}, \tag{1.16}$$

and $m^{i+1} = m^{i+k+1}$, which however is impossible in view of the second claim of the previous lemma.

□

Remark 1.2.1. There exist infinite non converging condensing sequences: Consider a simultaneously condensing sequence with two mass points $m_n = m_s$, where m_n is the mass of a point in the north pole of unit circle and m_s is a mass of a point in the south pole. Note that, this metric space is not a finite metric space but to explain this example in a finite metric space, one can use only four points metric spaces, namely the north, the south pole and the midpopints of them on the unit circle. Here, m is given as

$$m := m_s \delta_{\frac{3\pi}{2}} + m_n \delta_{\frac{\pi}{2}}.$$

1.2. Convergence

If we consider the rule of simultaneously moves (Hegselmann-Krause) studied by [17]. An admissible moves scenario is

1. **Step 1.** m_n moves to π and m_s moves to 0. Hence,

$$f_1(m)(\frac{\pi}{2}) = f_1(m)(\frac{3\pi}{2}) = 0 \quad \text{and} \quad f_1(m)(0) = m_s \quad \text{and} \quad f_1(m)(\pi) = m_n.$$

2. **Step 2.** m_n moves to $\frac{3\pi}{2}$ and m_s moves to $\frac{\pi}{2}$. Hence,

$$f_2(f_1(m))(0) = f_2(f_1(m))(\pi) = 0$$

and

$$f_2(f_1(m))(\frac{\pi}{2}) = m_s \quad \text{and} \quad f_2(f_1(\frac{3\pi}{2}) = m_n.$$

3. **Step 3.** The same as step 1.

4. **Step 4.** The same as step 2.

The condensing sequence constructed above $m^1, m^2, \ldots$ is simultaneously condensing and does not converge. We can also construct another type of non converging condensing sequences. We believe that, in this case, non converging sequences have a periodic behavior.

Example 1.2.1. Let us consider a two points metric space $\{x_1, x_2\}$ and two masses m_1 and m_2 in x_1 and x_2 respectively. Assume that, the masses are not necessary equal and they are not isolated. Let us suppose a simultaneously condensing of

$$m = m_1\delta_{x_1} + m_2\delta_{x_2},$$

Step 1. m_1 moves to m_2, while $E((x_1, x_2, m)) = 0$.
And m_2 moves to m_1, while $E((x_2, x_1, m)) = 0$. set

$$m^2 = m_2\delta_{x_1} + m_1\delta_{x_2}.$$

Step 2. m_2 moves to m_1, while $E((x_2, x_1, m)) = 0$.

And m_1 moves to m_2, while $E((x_1, x_2, m)) = 0$.

set
$$m^3 = m_1 \delta_{x_1} + m_2 \delta_{x_2}.$$

Step 3. The same as step 1.

If $m^1, m^2, \ldots$ represents this condensing sequence, then it not converges. In this case, the sequence is periodic.

1.3 Random metric on finite metric space

Our motivation for the construction of random metric is to try to simulate many numerical problems on finite metric spaces with respect to an arbitrary metric. Therefore, we present a method to construct random metrics in a finite metric space. Metrics of n points metric space are characterized by specified axioms. The split-metrics, used in [2, 11, 12] (also called cut metrics), are elements of the extremal rays on the cone of pseudometrics, and note that there are 2^{n-1} split-metrics on a n point set, for more detailed analysis, we refer to works of Dress and al. [12].

1.3.1 Extremal pseudometrics

Let us recall the definition of pseudometric. A map $d : X \times X \to \mathbb{R}^+$ is called pseudometric, if d satisfies the following conditions:

(i) $d(x, y) = 0$,

(ii) $d(x, y) = d(y, x)$, $\forall x, y \in X$,

(iii) $d(x, y) \leq d(x, z) + d(z, y)$, $\forall x, y, z \in X$.

1.3. Random metric on finite metric space

Let us denote by M_0 the set of all pseudometrics on X (n points set). For a given set X, a pseudometric d is called extremal if for all $g, h \in M_0$ the following holds

$$g + h = d \quad \text{implies} \quad g = \nu d, \; h = \mu d \quad \text{with} \quad \nu, \mu \geq 0.$$

a cut or split metric is a pseudometric given as:

$$d(x,y) := \begin{cases} 0, & \text{if } (x,y) \in A \times A \text{ or } A^c \times A^c \\ 1, & \text{otherwise.} \end{cases} \quad (1.17)$$

where A is a subset of X and A^c its complement in X. The cut metrics are extremal in the cone of all pseudometrics on a finite metric space: The proof can be found in the works of Dress and al. [12, 2, 11]. We denote by EX_n (resp. EX) the set of all cut metrics (resp. the set of all extremal metrics) in a n points metric space. These are given as:

$$EX_n := \{c_n^i | i = 1, \ldots, 2^{n-1}\} \subset EX := \{e | e \; extremal \; pseudometric\}, \quad (1.18)$$

where c_n^i is a cut metric and c_n is an extremal pseudometric. The cardinal of EX_n is exactly 2^{n-1} but the cardinal of EX is until now unknown. In order to construct a random metric, we use the following lemmas:

Lemma 1.3.1. Let X be a n points set. For $\lambda_e \in \mathbb{R}^+$, for all $e \in EX$, $\alpha, \beta \in X$, the map d is defined as:

$$d(\alpha, \beta) := \sum_{e \in EX} \lambda_e e(\alpha, \beta), \quad (1.19)$$

is a pseudometric and if $\lambda_e > 0$ for all $e \in EX$, then d is a metric.

Proof. Since the positive scalar multiplication with a pseudometric is a pseudometric and the sum of pseudometrics is a pseudometric, the sum (1.19) is a pseudometric. Also note that the set of split metrics is a subset of all extremal metrics.

1.3. Random metric on finite metric space

Now denote and suppose that $\lambda_i > 0$ for $i = 1\ldots, 2^{n-1}$, for all $\alpha, \beta \in X$, the set $\{\alpha\}$ is a subset of X and $X \setminus \{\alpha\}$ is the complement set of $\{\alpha\}$. Since $\alpha \neq \beta$, it follows that $\beta \in X \setminus \{\alpha\}$, define according to "theorem" (1.17) the following extremal pseudometric

$$d(x,y) := \begin{cases} 0, & \text{if } (x,y) \in \{\alpha\} \times \{\alpha\} \text{ or } X \setminus \{\alpha\} \times X \setminus \{\alpha\} \\ 1, & \text{otherwise.} \end{cases} \qquad (1.20)$$

d is an element of EX_n, so there exists i such that

$$c_n^i(\alpha, \beta) = 1.$$

Hence for all $\alpha, \beta \in X$ exists c_n^i such that

$$c_n^i(\alpha, \beta) > 0,$$

and if $\lambda_i > 0$ then $d(\alpha, \beta) > 0$, which means that d is metric.

□

Definition 1.3.1. Under the notation of Lemma 1.3.1, d is called random metric, if the choice of λ_e and/or the choice of e is random.

For our numerical simulation of random metrics, we use the following lemmas to construct a class of random metrics from the class of cut metrics:

Lemma 1.3.2. Let (X, d) be a n points pseudometric space. If λ_i, for $i = 1\ldots, 2^{n-1}$, is a sequence of independent and identically distributed random variable on a probability space $(\Omega, \mathcal{F}, P)$ with realization in $]0, \infty[$, then the map $d(\cdot, \cdot)(\omega)$ defined as

$$d(\alpha, \beta)(\omega) := \sum_{i=1}^{2^{n-1}} \lambda_i(\omega) c_n^i(\alpha, \beta), \quad c_n^i \in EX_n, \qquad (1.21)$$

is a class of random metrics extracted from the set of the cut metrics.

1.3. Random metric on finite metric space

Proof. Since each realization $\lambda_i(\omega)$ is strictly positive, the proof follows immediately from lemma 1.19.

□

Lemma 1.3.3. Let (X, d) a n points pseudometric space. For $I \subset \{1 \ldots, 2^{n-1}\}$, $\lambda_i \in \mathbb{R}^+$ for $i \in I$, and $\lambda > 0$, the map d defined as

$$d(\alpha, \beta) := \lambda \delta(\alpha, \beta) + \sum_{i \in I} \lambda_i c_n^i(\alpha, \beta), \quad c_n^i \in EX_n, \tag{1.22}$$

is a pseudometric and if λ is strictly positive, d is a metric. Here, δ is the discrete metric given as:

$$\delta(x, y) := \begin{cases} 1, & \text{if } x \neq y \\ 0, & \text{otherwise.} \end{cases} \tag{1.23}$$

Proof. Since the sum of pseudometrics is a pseudometric for all λ_i and for all λ the sum (1.22) is a pseudometric. If $\lambda > 0$, then for all $\alpha, \beta \in X$ the distance $d(\alpha, \beta)$ is strict positive since $\lambda(\alpha, \beta)$ is strictly positive.

□

Lemma 1.3.4. Let (X, d) be a n points pseudometric space. For $p \in X$, the following pseudometric is extremal

$$d_p(x, y) := \begin{cases} 1, & \text{if } (x = p \text{ and } y \neq p) \text{ or } (x \neq p \text{ and } y = p) \\ 0, & \text{otherwise.} \end{cases} \tag{1.24}$$

Proof. Trivial.

Theorem 1.3.1. Let $X := \{a_1, \ldots, a_n\}$ be a finite set of points. Any metric d defined on X, solution of $C_n^2 - 1$ linearly independent equations of the form

$$d(a_i, a_j) = 0 \quad \text{and or} \quad d(a_t, a_r) = d(a_t, a_s) + d(a_s, a_r), \tag{1.25}$$

for $i, j, k, s, r \in \{1, \ldots, n\}$, is extremal.

Proof. See [12].

□

Lemma 1.3.5. Any metric d can be written as a convex combination of an extremal metric and a metric:

$$d = \alpha d_e + (1-\alpha)d', \qquad (1.26)$$

where d_e is an extremal metric, d' is a metric and $\alpha \in [0,1]$.

Proof. Trivial.

Remark 1.3.1. The construction of a random metric is based on:

(i) random choice of the coefficients λ_e of lemma 1.3.1 or

(ii) random choice of the linearly independent hyperplane equations or

(iii) all assumptions together.

For our numerical simulations of random numbers, we use two random variables, the first one is uniformly distributed in any positive set of $I\!R^+$, the second one is normally distributed. The construction of the normal distribution is done by using the well known Box-Muller method, for more details see [25]. The metric is then constructed using one of the algorithms presented bellow.

1.3.2 Algorithms for the construction of random metrics

In this section we present some algorithms to construct metrics and random metrics.

Algorithm 1. From lemma 1.3.4 one can extract the following algorithm to construct a class[1] of random metrics

---
[1]This class of random metrics is based on the split-metrics.

1. Set $d=0$ zero-metric and $p=1$.
2. Define d_p as ,
$$d_p(i,j) := \begin{cases} 1, & \text{if } (i=p; j \neq p) \text{ or } (i \neq p; j=p) \\ 0, & \text{otherwise.} \end{cases}$$
3. Chose randomly $a_p > 0$;
   Set $d = d + a_p d_p$,
4. if $p=n$, return d; STOP:
   else Set $p = p+1$, go to 2.
   End.

Algorithm 2. From lemma 1.3.3, it follows the following this algorithm for construction only a class of random metrics extracted from split metrics:

1. Set $d=0$ zero-metric, $p=1$ and $j=1$.
2. Define d_p a cut metric of $\{1,\ldots,n\}$ as ,
3. For $p=1$ to 2^{n-1}
   construct the cut metrics d_p
4. For $j=1$ to 2^{n-1} do
   Chose randomly $a_j > 0$;
   Set $d = d + a_j d_j$,
   return d, End.

Algorithm 3. Here is another algorithm to construct random metrics from the same class of split metric as in the preceding algorithm:

1. Set $d=0$ zero-metric, M large integer.
2. For $j=1$ to M
   Chose randomly a cut metric d_j and $a_j > 0$;

1.3. Random metric on finite metric space

Set $d = d + a_j d_j$,

3. Chose randomly $a > 0$;

    Set $d = a + d$,

    return d, End.

Algorithm 4. More generally, using theorem 1.3.1 and lemma 1.26, the following algorithm can be used to construct any random metric from a convex combination of extremal pseudometrics:

1. Let $s(d) = \text{Sum } d(x,y)$, Set $d = 0$ zero-metric and p=0.
2. Chose randomly $C_n^2 - 1$ equation of type:

    $d(x,y) = 0$ and/or $d(x,y) = d(x,z) + d(z,y)$,

    solve for d' with $s(d') = 1$;
3. For a random number $a \in (0,1)$ Set $d = ad + (1-a)d'$;

    and p=p+1.
4. if $p = C_n^2 + 1$, return d; STOP:

    else go to 2.

    End.

Example 1.3.1. For numerical constructions of random metrics, we have used uniform and normal distributions of random numbers, see [25]. In order to compare a random metric to an Euclidean metric, we also define in matrix form the Euclidean and random metrics matrices. The matrix d_e, d_u and d_n represent respectively metric matrix of the Euclidean, the random metric with uniform distributed random entries and the random metric with normal distributed random entries. In order to compute the Euclidean metric of a finite set of points, let us

1.3. Random metric on finite metric space

define X as:

$$X := \{0, 0.1, 0.2, 0.3, 0.4, 0.5, 0.6, 0.7, 0.8, 0.9, 1\} \subset [0,1].$$

The following matrix is the Euclidean metric of the set X:

$$d_e = \begin{pmatrix} 0 & 0.1 & 0.2 & 0.3 & 0.4 & 0.5 & 0.6 & 0.7 & 0.8 & 0.9 & 1 \\ 0.1 & 0 & 0.1 & 0.2 & 0.3 & 0.4 & 0.5 & 0.6 & 0.7 & 0.8 & 0.9 \\ 0.2 & 0.1 & 0 & 0.1 & 0.2 & 0.3 & 0.4 & 0.5 & 0.6 & 0.7 & 0.8 \\ 0.3 & 0.2 & 0.1 & 0 & 0.1 & 0.2 & 0.3 & 0.4 & 0.5 & 0.6 & 0.7 \\ 0.4 & 0.3 & 0.2 & 0.1 & 0 & 0.1 & 0.2 & 0.3 & 0.4 & 0.5 & 0.6 \\ 0.5 & 0.4 & 0.3 & 0.2 & 0.1 & 0 & 0.1 & 0.2 & 0.3 & 0.4 & 0.5 \\ 0.6 & 0.5 & 0.4 & 0.3 & 0.2 & 0.1 & 0 & 0.1 & 0.2 & 0.3 & 0.4 \\ 0.7 & 0.6 & 0.5 & 0.4 & 0.3 & 0.2 & 0.1 & 0 & 0.1 & 0.2 & 0.3 \\ 0.8 & 0.7 & 0.6 & 0.5 & 0.4 & 0.3 & 0.2 & 0.1 & 0 & 0.1 & 0.2 \\ 0.9 & 0.8 & 0.7 & 0.6 & 0.5 & 0.4 & 0.3 & 0.2 & 0.1 & 0 & 0.1 \\ 1 & 0.9 & 0.8 & 0.7 & 0.6 & 0.5 & 0.4 & 0.3 & 0.2 & 0.1 & 0 \end{pmatrix}$$

For an 8 points metric space and using a uniform distribution of positive random numbers, we generate the matrix d_u, which is a metric:

$$d_u = \begin{pmatrix} 0 & 0.5062 & 0.4627 & 0.5693 & 0.2864 & 0.4828 & 0.3735 & 0.3238 \\ 0.5062 & 0 & 0.634 & 0.7406 & 0.4577 & 0.654 & 0.5447 & 0.4951 \\ 0.4627 & 0.634 & 0 & 0.6971 & 0.4142 & 0.6106 & 0.5013 & 0.4516 \\ 0.5693 & 0.7406 & 0.6971 & 0 & 0.5208 & 0.7172 & 0.6079 & 0.5582 \\ 0.2864 & 0.4577 & 0.4142 & 0.5208 & 0 & 0.4343 & 0.325 & 0.2753 \\ 0.4828 & 0.654 & 0.6106 & 0.7172 & 0.4343 & 0 & 0.5214 & 0.4717 \\ 0.3735 & 0.5447 & 0.5013 & 0.6079 & 0.325 & 0.5214 & 0 & 0.3624 \\ 0.3238 & 0.4951 & 0.4516 & 0.5582 & 0.2753 & 0.4717 & 0.3624 & 0 \end{pmatrix}$$

For an 8 points metric space and using a normal distribution of positive random numbers, we generate the matrix d_n, which also is a metric:

$$d_n = \begin{pmatrix} 0 & 0.4717 & 0.5044 & 0.4676 & 0.5394 & 0.5018 & 0.5124 & 0.5446 \\ 0.4717 & 0 & 0.4621 & 0.4253 & 0.4971 & 0.4595 & 0.4701 & 0.5023 \\ 0.5044 & 0.4621 & 0 & 0.458 & 0.5298 & 0.4923 & 0.5029 & 0.5351 \\ 0.4676 & 0.4253 & 0.458 & 0 & 0.493 & 0.4554 & 0.466 & 0.4982 \\ 0.5394 & 0.4971 & 0.5298 & 0.493 & 0 & 0.5272 & 0.5378 & 0.57 \\ 0.5018 & 0.4595 & 0.4923 & 0.4554 & 0.5272 & 0 & 0.5003 & 0.5325 \\ 0.5124 & 0.4701 & 0.5029 & 0.466 & 0.5378 & 0.5003 & 0 & 0.543 \\ 0.5446 & 0.5023 & 0.5351 & 0.4982 & 0.57 & 0.5325 & 0.543 & 0 \end{pmatrix}$$

In order to compare between the entries of metrics, we introduce the vector d_v given as:

$$d_v = (d_{1,2}, \ldots, d_{1,n}, d_{2,3}, \ldots, d_{2,n}, \ldots, d_{n-1,n}). \tag{1.27}$$

d_v is called metric vector. We denote d_v in our simulation by

$$x = \left(x_1, \ldots, x_{C_n^2}\right). \tag{1.28}$$

For NP=25 and NP=50, we plot respectively, the Euclidean metric vector of an equidistance discretization of the unit interval. The figure 1.1 gives a graphical idea of the Euclidean metric vector. We plot the metric vector of the Euclidean metric for NP=30 and NP=50 points metric space. In the following, we generate for each NP (NP=30 and NP=50: number of points), two simulations of random metrics: figure 1.2 presents eight simulation of random metric vectors, where for each NP, we run our code two times to get two simulations of a random metric of NP=30 points metric space and two simulations for NP=50. Let us denote by (Uniform) resp. (Normal) a random metric generated by using a uniform resp. normal distributed variable. It is important to note that the two dimensional plots of random metrics are only introduced to find a way to compare graphically random metrics with an Euclidean metric. Since the metric is only defined on the finite

1.3. Random metric on finite metric space

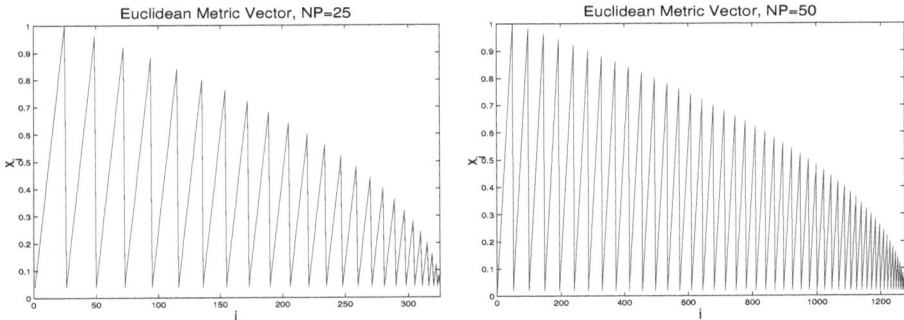

Figure 1.1: *Euclidean metric vector, NP=25, and 50.*

metric space, we interpolate the graph to get a continuous two dimensional surface. Also, it is important to note that using uniform or normal random numbers does not imply that the distribution of the metrics, as a random variable, is similar. This is not subject of our study.

Remark 1.3.2. Note that the Euclidean metric has the same behavior for NP=25 and NP=50, but the random metric vector has different behaviors, this is due to the choice of the number of points (NP=25 and NP=50) or to the random effect of the the random generator. We have also remarked that, the uniform random metric vector has high noise compared to the normal one.

1.3.3 Numerical simulations of random metrics

In this section, we present the two dimensional mesh of numerical simulation of random metrics and the Euclidean on the unit interval. For NP=25, 50 grid points and using the uniform and the normal distributed random numbers, we generate for each NP two different simulations of random metrics. Therefore, we plot the two dimensional function $(x, y) \mapsto d_k(x, y)$ with $k = e, u, n$, where the abbreviation e, u, n represent respectively the Euclidean, uniform random distribution

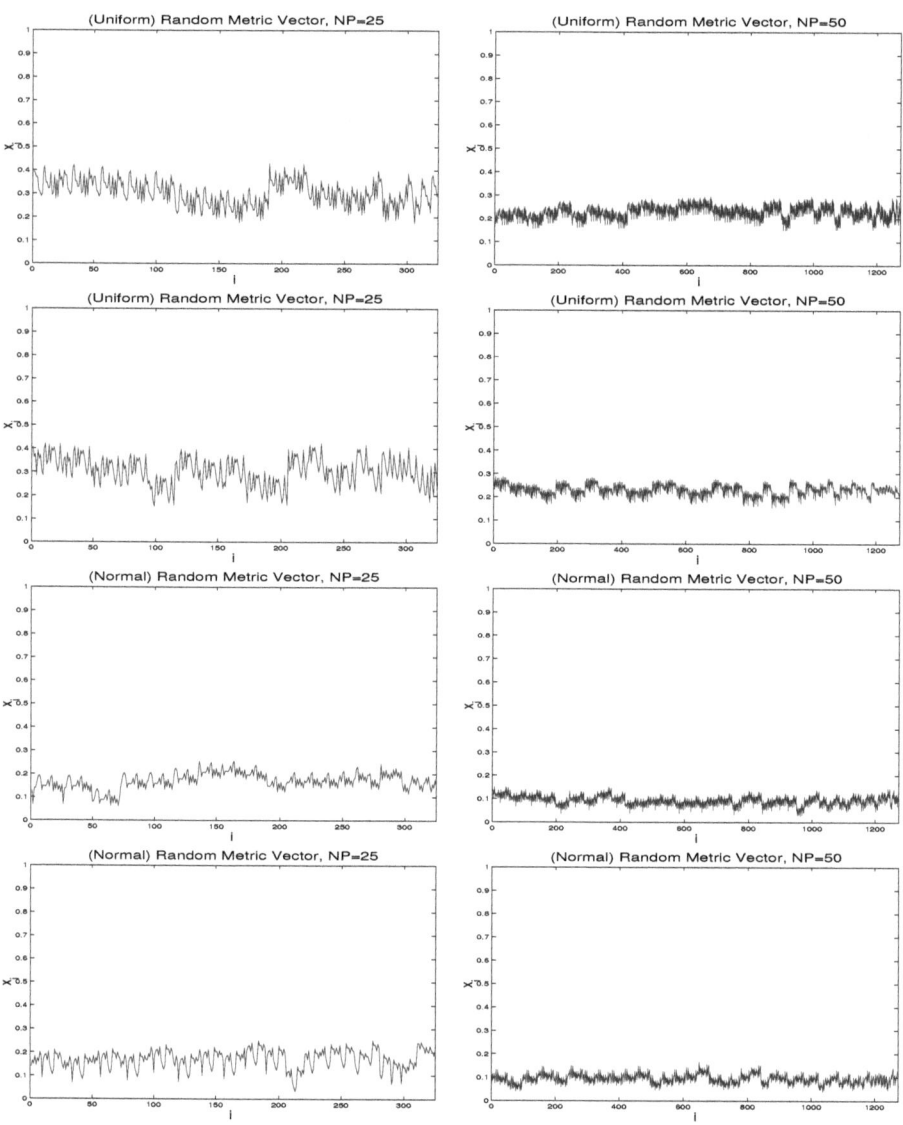

Figure 1.2: *Random metric vector (Uniform, Normal), NP=25, and 50.*

1.3. Random metric on finite metric space

and normal random distribution. In figure 1.3, we plot the Euclidian metric mesh

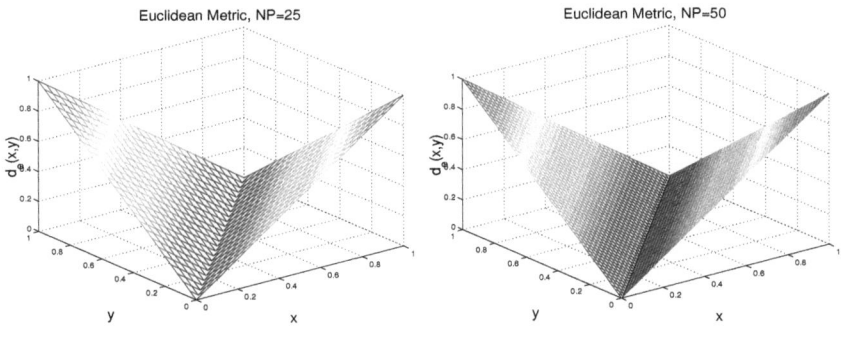

Figure 1.3: *Euclidean metric mesh, NP=25, and 50.*

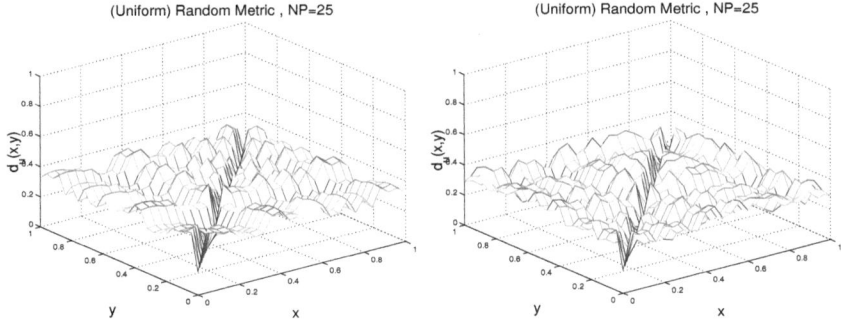

Figure 1.4: *Random metric mesh (Uniform), NP=25.*

of the unit interval for NP=25 and NP=50 grid points. In figure 1.4 and 1.5, we generate, using a uniform random number generator, random metrics for different number of points NP=25 and NP=50 respectively. The figure 1.6 presents for each NP two simulation of random metrics using a uniform and a normal distribution of random numbers.

1.3. Random metric on finite metric space

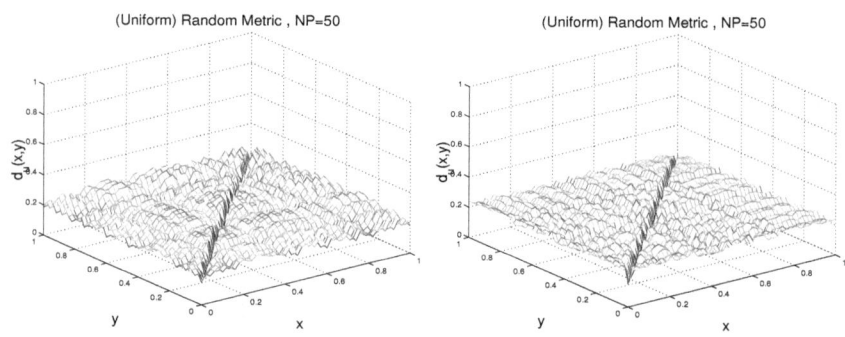

Figure 1.5: *Random metric mesh (Uniform), NP=50*.

The figures 1.7 and 1.8 show the contour of the plots in 1.6 and 1.6 respectively for the (normal) random metric for NP=50. The symmetry of the metric is clearly shown. To be adapted to the matrix form of the two dimensional of a matrix, we have turned the plot in this form.

We remark that, the Euclidean metric is smooth and totaly different from the other generated metrics. In order to show the symmetry of a random metric, we have plotted the contours in figure 1.7 and figure 1.8 of four simulations. These figures present respectively the symmetric behavior of random metrics.

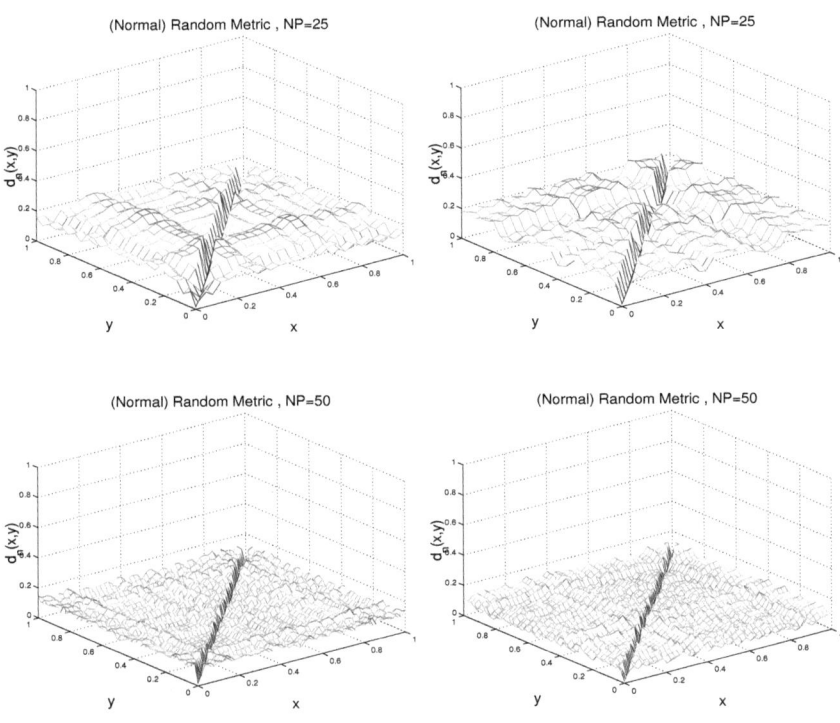

Figure 1.6: *Random metric mesh (Normal), NP=25, and NP=50.*

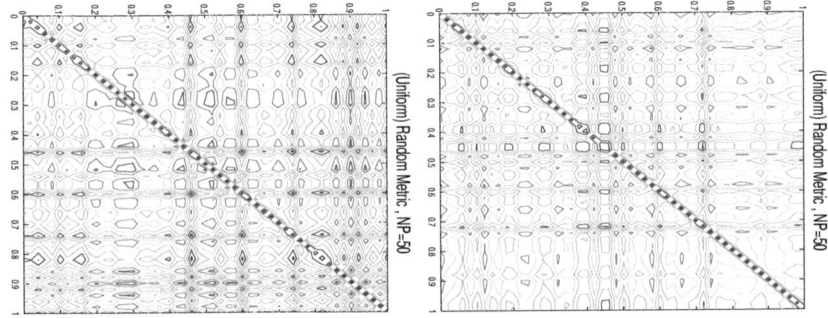

Figure 1.7: *(Uniform) Random metric contour, NP=50.*

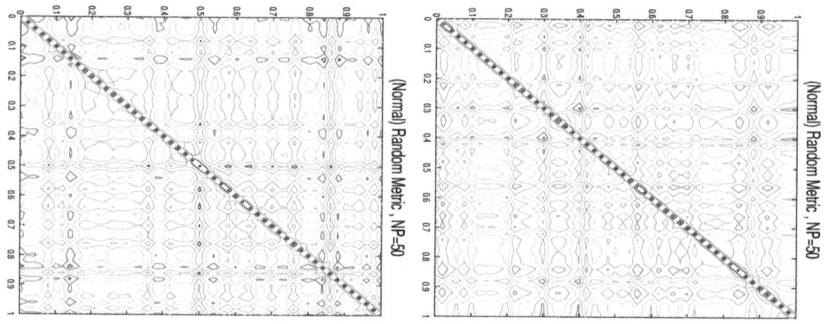

Figure 1.8: *(Normal) Random metric contour NP=50*.

1.4 Numerical simulations of condensing sequences

In our simulations, we do numerical experiments on three finite metric spaces. Each finite space will be constructed as n points metric space and a subset of a continuous metric space. The first one is a finite subset of the real plane, the second one is a finite subset of a one dimensional manifold, to simplify we use in this case an approximation of the unit circle. The numerical simulations are listed as follows:

1. **Simulations on an Euclidean finite metric space:**

   (a) Uniform mass distribution on the FMS.

   (b) Uniform mass distribution on the FMS.

   (c) Uniform random mass distribution on the FMS.

2. **Simulations on circular finite metric space:**

   (a) Uniform mass distribution.

   (b) Uniform mass distribution.

   (c) Uniform random mass distribution.

3. **Simulations with respect to a random metric constructed from the class of split metrics:**

   (a) Simulation respect to a random metric.

   (b) Simulation respect to a random metric.

   (c) Simulation respect to a random metric.

1.4.1 Simulations on an Euclidean finite metric space

This section propose simulations on a finite subset of an Euclidean finite metric space. We define a finite metric space of 121 points (X, d) as:

$$X = \{x_1, \ldots, x_{121}\} \subset \mathbb{R}^2, \tag{1.29}$$
$$d(x_i, x_j) = \|x_i - x_j\|_2, \forall i, j = 1, \ldots, 121,$$

where since $X \subset \mathbb{R}^2$, the metric used here is the Euclidean one. The initial measure will defined as a positive measure given as $m := \sum_{x \in X} m(x) \delta_x$, where $S(m) = X$ and $m(x) > 0$. We run our code after fixing a random order of reactions (the array of 121 index will randomly permuted). It is important to note that the positions, which minimize the energy are not unique, therefore we choose randomly one of them. Note that the uniform random distribution generate real random numbers between 0 and 4. The following table summarizes the results of the simulations on the Euclidean finite metric space:

Remark on the results 1.4.1. We have observed that the limit states of simulations (a), (b) and (c)

1. have different distribution of mass,
2. are ε isolated groups,
3. have a non uniform distribution of mass.

1.4. Numerical simulations of condensing sequences

Parameter/Sim.	(a)	(b)	(c)
NP	121	121	121
ε	0.19	0.19	0.19
Initial state	121 masses (one)	121 masses (one)	121 masses (in U(0,4))
Final state	21 isolated masses	27 isolated masses	21 isolated masses
Time in Sec	190	168	262
Number of iterations	300	300	300

Table 1.1: Results of simulations in an (Euclidean) finite metric space.

In matrix form, we give the distribution of masses for the limit measure of simulations (a). Note that, the entries of the matrix give the number of particles in each point of the 121 points space: (similar for (b)).

$$\begin{bmatrix} 0 & 3 & 0 & 1 & 0 & 0 & 0 & 3 & 0 & 0 & 0 \\ 0 & 0 & 0 & 0 & 0 & 5 & 0 & 0 & 0 & 0 & 3 \\ 0 & 0 & 0 & 0 & 0 & 0 & 0 & 0 & 0 & 0 & 0 \\ 0 & 0 & 4 & 0 & 0 & 0 & 0 & 7 & 0 & 0 & 0 \\ 4 & 0 & 0 & 0 & 8 & 0 & 2 & 0 & 0 & 0 & 0 \\ 0 & 0 & 4 & 0 & 0 & 0 & 0 & 6 & 0 & 0 & 0 \\ 0 & 0 & 0 & 0 & 9 & 0 & 0 & 0 & 0 & 0 & 0 \\ 0 & 0 & 0 & 0 & 0 & 7 & 0 & 0 & 0 & 0 & 0 \\ 9 & 0 & 0 & 0 & 0 & 0 & 0 & 0 & 9 & 0 & 0 \\ 0 & 0 & 0 & 3 & 0 & 0 & 0 & 8 & 0 & 0 & 0 \\ 0 & 13 & 0 & 0 & 0 & 9 & 0 & 0 & 0 & 0 & 4 \end{bmatrix},$$

and also for simulation (c), the following matrix represent the distribution of mass at the limit state. Note that, the entries of the matrix give the sum of the mass

1.4. Numerical simulations of condensing sequences

of the particles in each point of the 121 points space:

$$\begin{bmatrix} 0 & 0 & 0 & 0 & 0 & 0 & 0 & 0 & 5.24 & 0 & 0 \\ 0 & 0 & 13.59 & 0 & 0 & 0 & 13.52 & 0 & 0 & 0 & 0 \\ 14.91 & 0 & 0 & 0 & 0 & 0 & 0 & 0 & 0 & 9.12 & 0 \\ 0 & 0 & 0 & 0 & 13.36 & 0 & 0 & 19.49 & 0 & 0 & 0 \\ 0 & 0 & 7.66 & 0 & 0 & 0 & 0 & 0 & 0 & 0 & 8.41 \\ 9.52 & 0 & 0 & 0 & 7.62 & 0 & 0 & 0 & 0 & 0 & 0 \\ 0 & 0 & 0 & 0 & 0 & 0 & 0 & 18.93 & 0 & 0 & 9.00 \\ 0 & 22.06 & 0 & 0 & 0 & 10.58 & 0 & 0 & 0 & 0 & 0 \\ 0 & 0 & 0 & 15.03 & 0 & 0 & 0 & 0 & 13.73 & 0 & 0 \\ 0 & 0 & 0 & 0 & 0 & 0 & 0 & 0 & 0 & 0 & 0 \\ 29.62 & 0 & 0 & 0 & 24.56 & 0 & 17.59 & 0 & 0 & 16.61 & 0 \end{bmatrix}$$

The figures 1.9 present four condensing iterations in a finite metric space. We show the initial state, the iterations 50, 100, 150 and 300. The small dark dots represent the metric space and the large ones represent the particles. The initial measure is a collection of point masses such that each point of the grid has a mass one. A move is only admissible on the small points (FMS). In this case the limit measure is a collection of ε isolated mass points. We have plotted only 50 steps. Therefore, the limit state is not necessary the iteration 300. In this plot, we show the center of mass of each point mass, the weight is given in matrix form above. The last figure present the new repartition and the density of the particles at the limit state.

The figures 1.10 present five condensing iterations in a finite metric space the initial state, the iterations 50, 100, 150, and 300. Using the same initial data of simulation (a), the limit measure is a collection of ε isolated point masses. The last figure present the new repartition and the density of the particles at the limit state, where the used scale indicate the index of each nonnegative mass.

(a) Uniform and deterministic mass distribution on the FMS:

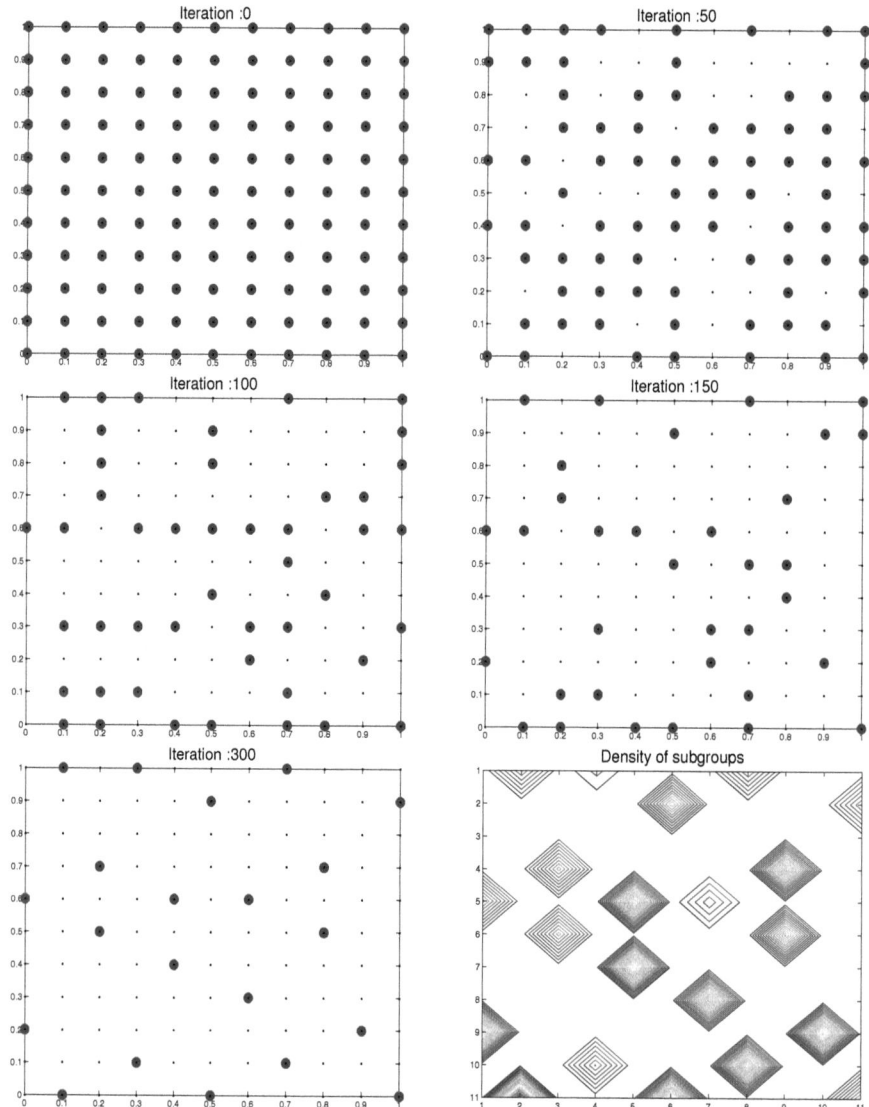

Figure 1.9: *Condensing in an Euclidean finite metric space (sim. (a)).*

(b) Uniform and deterministic mass distribution on the FMS:

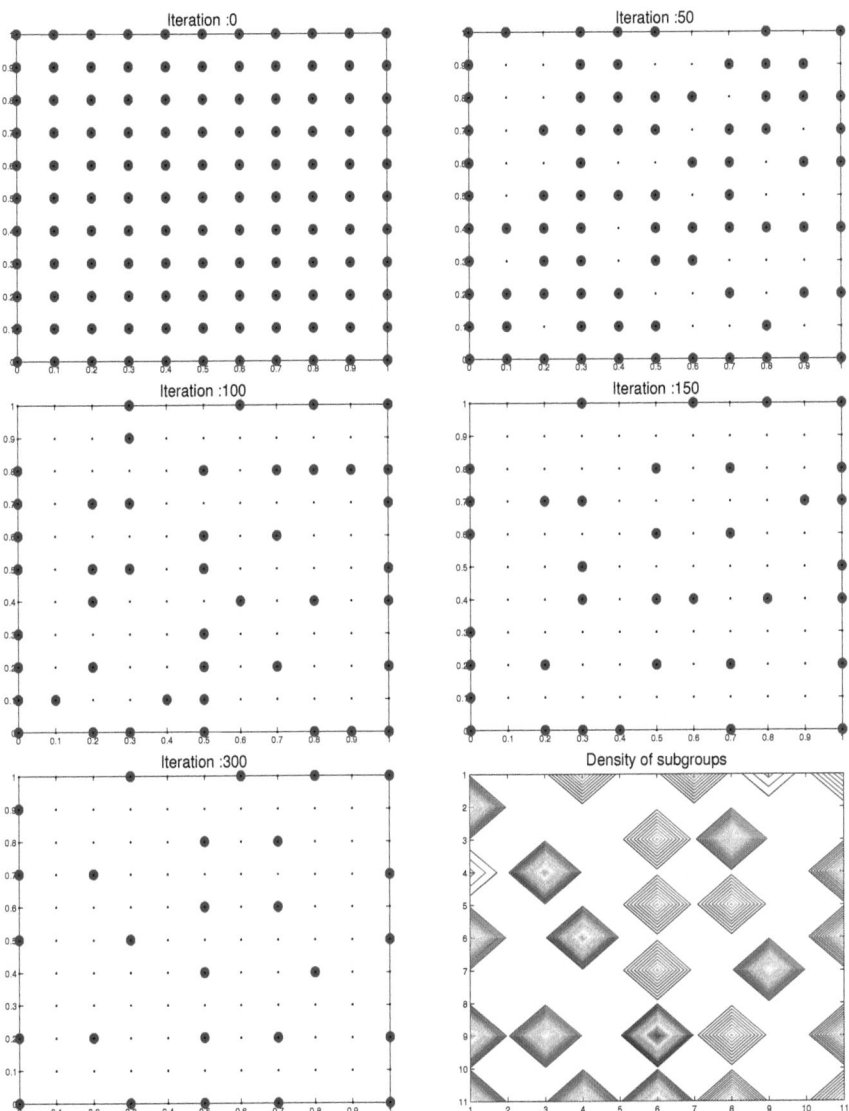

Figure 1.10: *Condensing in an Euclidean finite metric space (sim. (b))*

1.4. Numerical simulations of condensing sequences

(c) Uniform random mass distribution on the FMS:

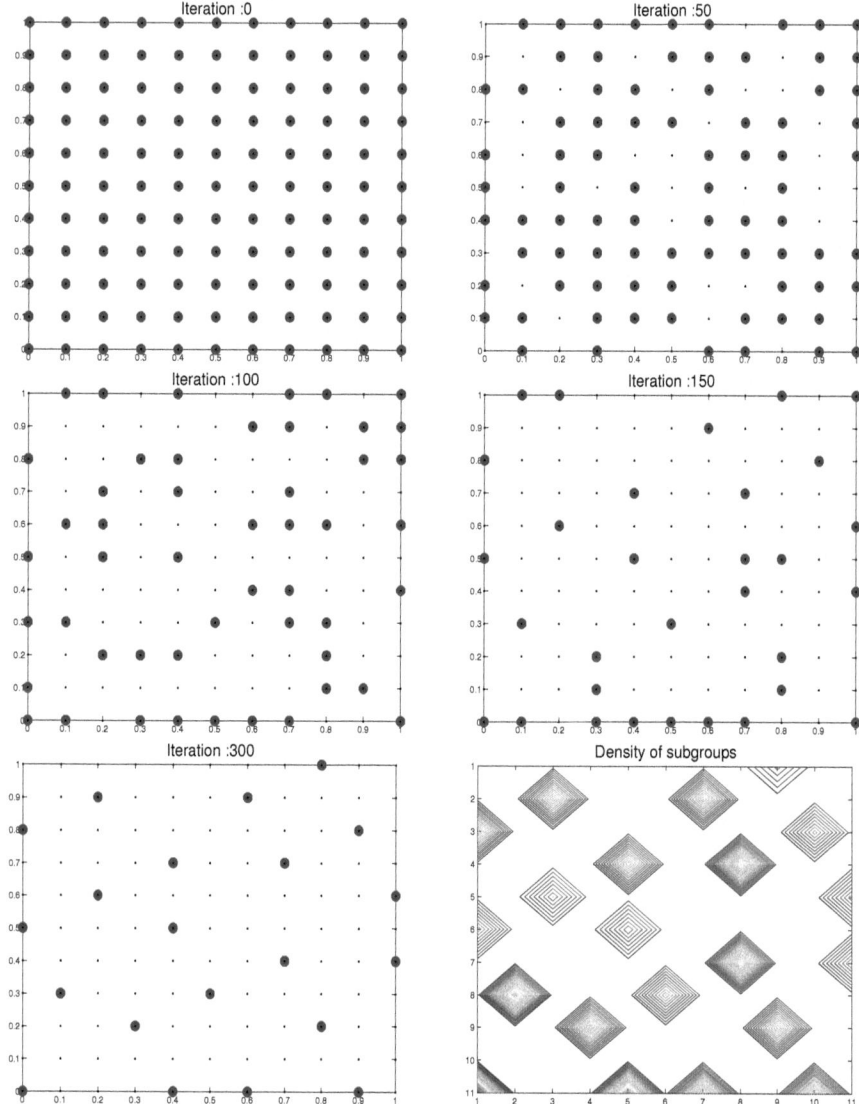

Figure 1.11: *Condensing in an Euclidean finite metric space (sim. (c))*.

1.4. Numerical simulations of condensing sequences

The figures 1.11 present three condensing iteration on a finite metric space. The initial measure is a discrete mass, in each points of the metric space, we generate a uniform random mass in $U(0,4)$. We show the initial measure and the iteration 100, 150 and 300. The limit measure (iter. 300) is constituted only of ε isolated point masses. In this plot, we show the center of mass of each point mass, the weight is given in matrix form above. The last figure present the new repartition and the density of the particles at the limit state, where the used scale indicate the index of each nonnegative mass.

1.4.2 Simulations in a finite circular metric space

This section do simulations in a finite circular metric space. We use an equidistance discritization of circular metric space. Let (X, g) be 100 points circular metric space given such that

$$X = \{x_1, \ldots, x_{100}\}, \tag{1.30}$$

$$d(x_i, x_j) = g(x_i, x_j), \forall i, j = 1, \ldots, 100.$$

Where the circular is defined as:

$$g(x_i, x_{i+1}) = 1,$$
$$g(x_i, x_{i+k}) = \min(k, 100-k), \quad \forall k = 1, \ldots, 100. \tag{1.31}$$

where the distance (1.31) is not an approximation of an arc even if our plots are done in a circle. We define the distance between two neighboring points points as a constant one. This can be understood as an approximation of the perimeter of a circle. The circle is used only to find a simple way to visualize our numerical results in a not Euclidean metric space. As in the previous simulations, the initial

1.4. Numerical simulations of condensing sequences

measure will be defined as a positive measure, such as in each point of space, we put a positive mass. m is given as $m := \sum_{x \in X} m(x)\delta_x$, where $S(m) = X$ and $m(x) > 0$. The exact choice of the masses is given in the table 1.4.2. After fixing a random order of reactions (the array of 100 index will randomly permuted), we run our code to simulate singular condensing of particles. We simulate the condensing process in three different times, (a), (b) and (c). The following table summarizes the results of the simulations:

Parameter/Sim.	(a)	(b)	(c)
NP	100	100	100
ε	2	2	2
Initial state	100 masses (one)	100 masses (one)	100 masses (in U(0,4))
Final state	27 isolated masses	28 isolated masses	26 isolated masses
Time in Sec	62	72	68
Number of iterations	200	180	190

Table 1.2: Results of simulations in a finite metric space (subset of S^1).

Remark on the results 1.4.2. The limit measures of condensing process in the geodesic metric space shows the same behavior. They are ε isolated point masses. For the simulation (a) and (b) and by using the same initial data the limit measures has the same behavior, but different distribution of mass. These describe stochastic condensing of particles. The output of simulation (c) has also the same characteristics, even if the initial point masses points are randomly generated with a uniform random distribution $U(0, 4)$. These results are summarizes in figure 1.15. Note also that, the order of actions and the position, which minimizes the energy are randomly chosen.

1.4. Numerical simulations of condensing sequences

(a) Simulations on circular finite metric space:

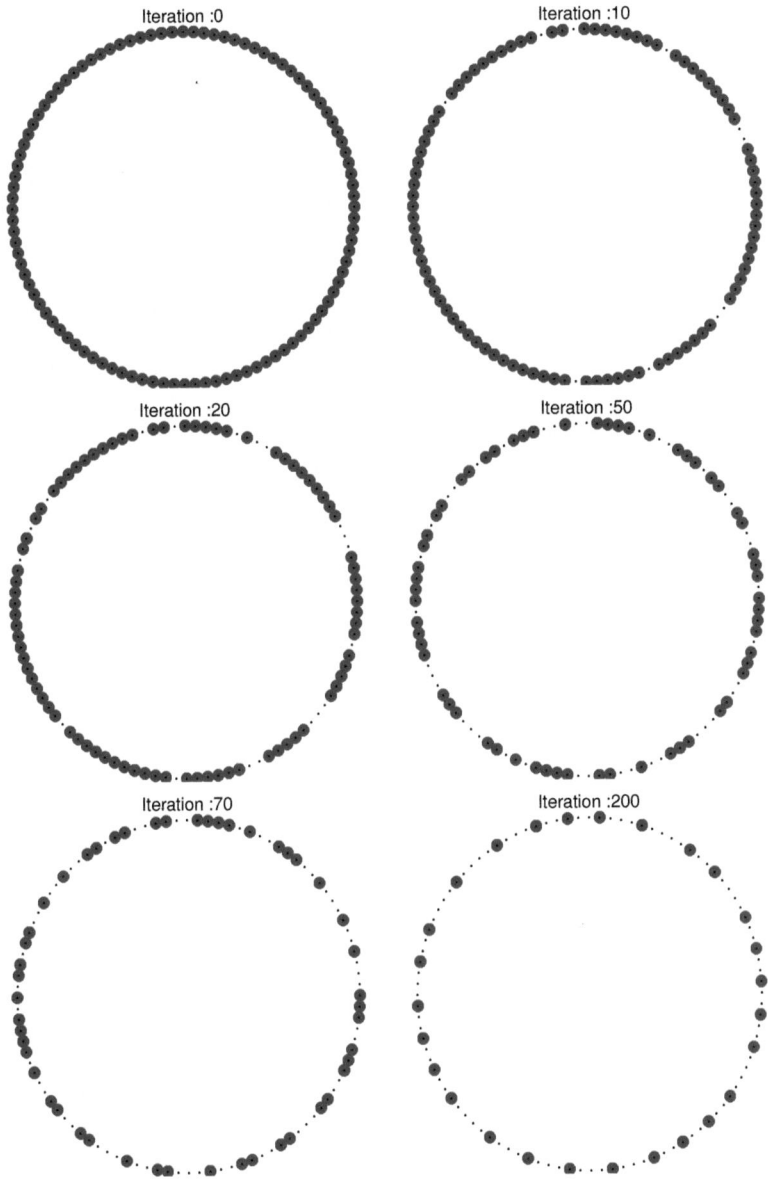

Figure 1.12: *Condensing in a geodesic finite metric space (sim. (a))*

(b) Simulations on circular finite metric space:

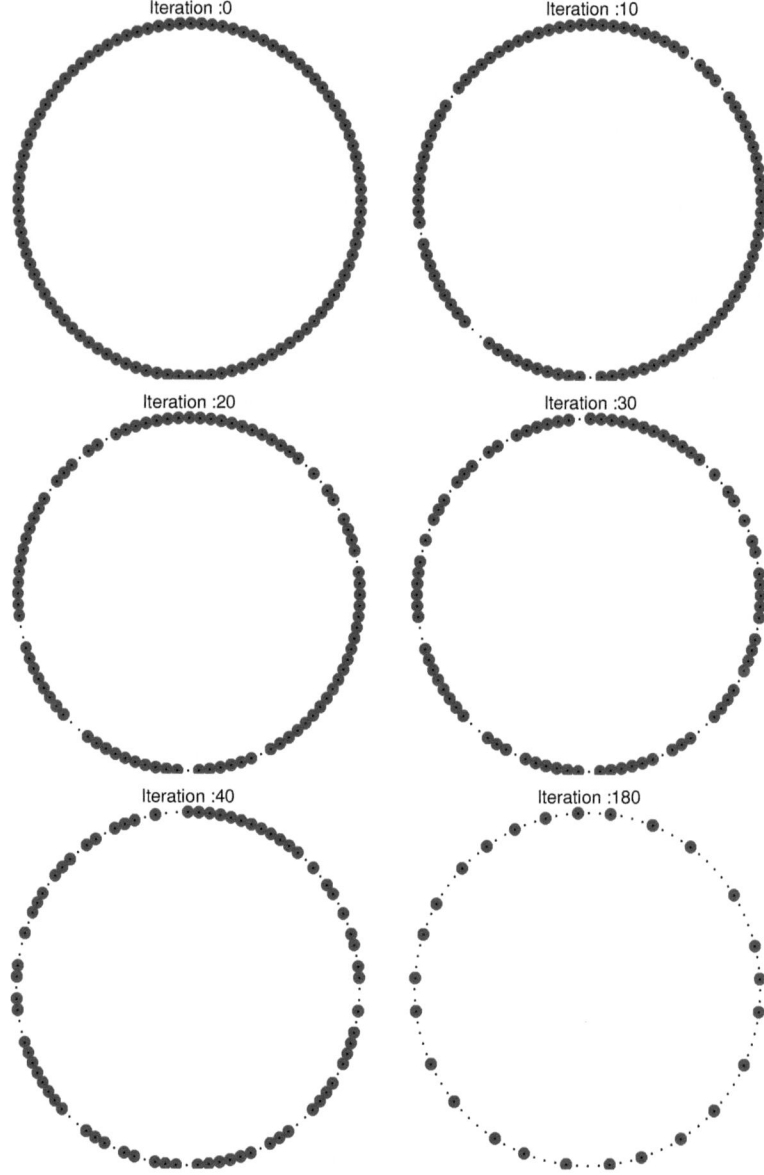

Figure 1.13: *Condensing in a geodesic finite metric space (sim. (b))*

mass.

1.4. Numerical simulations of condensing sequences

(c) Simulations on circular finite metric space:

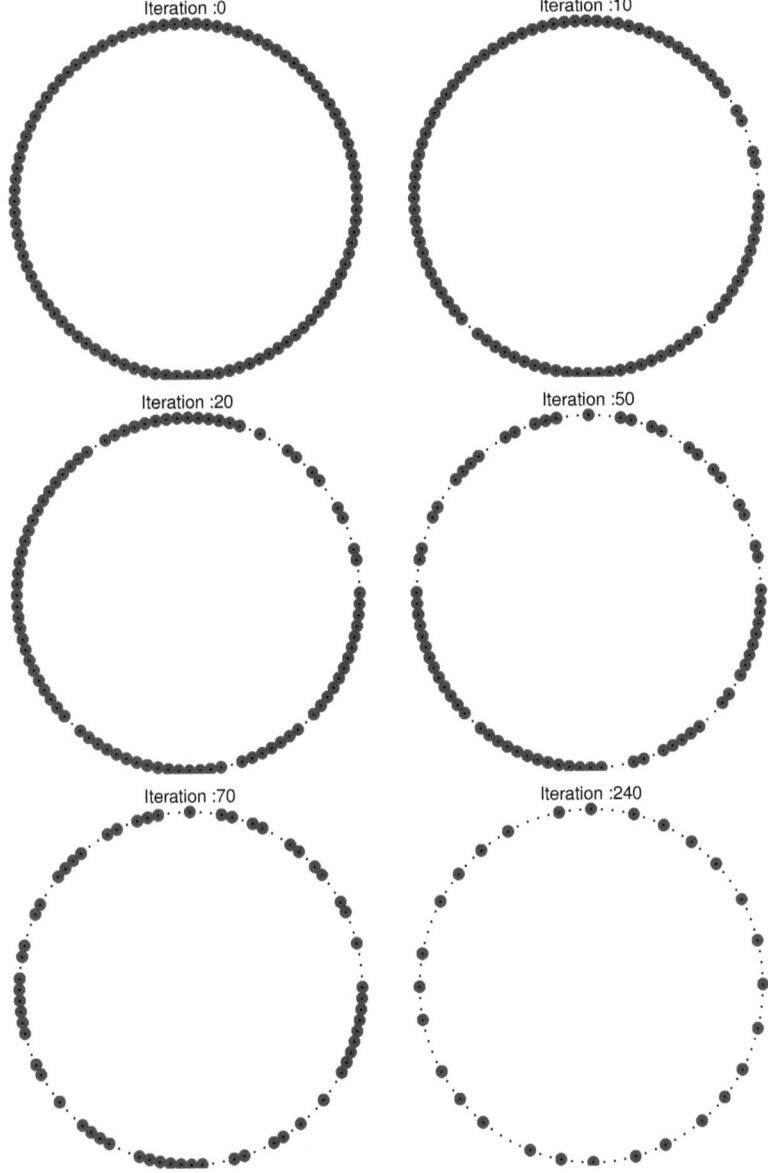

Figure 1.14: *Condensing in a geodesic finite metric space (sim. (c))*

1.4. Numerical simulations of condensing sequences

The figures 1.12 present six condensing iterations in a circular finite metric space. We show the initial state, the iterations 10, 20, 50, 70 and 200. The small dark dots represent the geodesic metric space and the large ones represent the particles. In each point of the metric space, we put a mass one. A move is only admissible on the small points (FMS). In this case the limit measure (iter. 200) is a collection of ε isolated point masses.

Figure 1.13: Using the same initial data of the preceding simulation, this figure presents six condensing iterations in a circular finite metric space, the initial state, the iterations 10, 20, 30, 40 and 180. In this case the limit measure (iter. 180) is a collection of ε isolated point masses.

Figure 1.14: Using the uniform random distribution $U(0,1)$, we generate the initial masses in each point of the FMS. We plot the initial state, the iterations 10, 20, 50, 70 and 240. In this case the limit measure (iter. 240) is a collection of ε isolated point masses. In this plot, we show the center of mass of each point mass.

The figures 1.15 show respectively the condensing process of our simulations on the finite circular metric space (sim. a, b and c), we also show the convergence behavior of the particles. The limit measures are an ε isolated point masses. Note that, we have plotted only the center of each point mass.

1.4. Numerical simulations of condensing sequences

Simulations on circular finite metric space (a), (b) and (c):

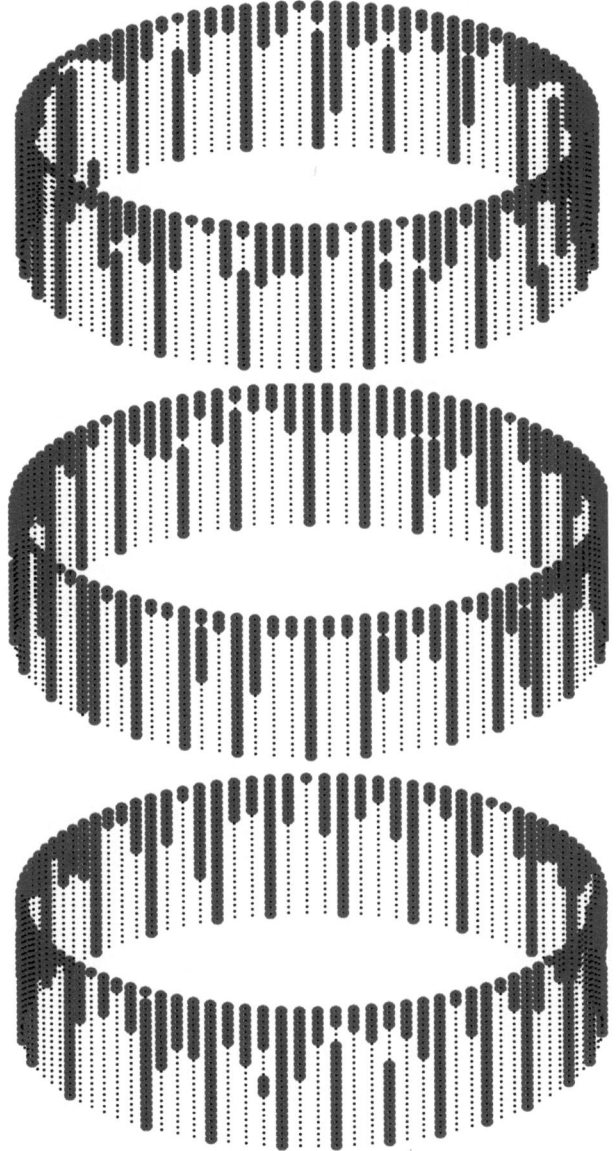

Figure 1.15: *Condensing in a geodesic finite metric space (sim. (a,b, c))*

1.4.3 Simulations with respect to a random metric

In this section, we apply random metrics. For such a construction of metric, we use the algorithm 3. Let us denote by (X, d_r) such as metric space, where d_r is a random metric:

$$\begin{aligned}
\text{define } x_i, x_j \text{ for } & i, j = 1, \ldots, 100 \text{ such that} \\
\text{distance } (x_i, x_j) \text{ is } & d_r(x_i, x_j), \forall i \neq j = 1, \ldots, 100, \\
d_r(x_i, x_i) = & 0, \forall i = 1, \ldots, 100, \\
X = & \{x_1, \ldots, x_{100}\},
\end{aligned} \quad (1.32)$$

where (X, d_r) is a finite metric space. It is important to note that the visualization of (X, d_r) given by (1.32) is not trivial, since the space is characterized only by its metric. In order to analyze condensing sequences with respect to a random metric, we define the so called position of a particle. We assume that each particle maintains a position at the start and in each one has a positive mass. We have also to note that two neighboring positions are independent from the length of the distance between them. This method is introduced only to find a way to visualize the condensing process with respect to a random metric. In our simulations the bar plot of the positions of particles, show in many time iteration, that the particles changes their positions with respect to the local energy rule (1.1.2). The axis of the bar plot presents the positions of the particles (x-axis) and the number of the particles in a given position (y-axis). Since the metric is unknown, the code chose a random $\varepsilon = d(x_1, x_5)$. The code breaks down if the energy is zero. We observe the condensing of a measure m given as $m := \Sigma_{x \in X} m(x) \delta_x$, where $S(m) = X$ and $m(x) > 0$. The following table summarizes the results of our simulation of condensing sequence with respect to a random metric:

1.4. Numerical simulations of condensing sequences

Parameter/Sim.	(a)	(b)	(c)
NP	100	100	100
ε	0.182508	0.200761	0.254162
Initial state	100 masses (one)	100 masses (one)	100 masses (one)
Final state	24 isolated masses	36 isolated masses	1 mass (consensus)
Run time	69 Sec	82 Sec	80 Sec
Number of iterations	102	180	136

Table 1.3: Results of simulations (random) finite metric space.

Remark on the results 1.4.3. The limit state even if there is only small difference between the ε have'nt similar behavior. Since in the third simulation the finale state is a consensus, the first and the second one are ε isolated masses. These results are also given by figures 1.19, 1.20 and 1.21, which show (respectively) that at the limit state all points with strictly positive masses have a distanced grater than ε between them, the decreasing number of the support and also the decreasing total energy.

The figures 1.16 present a condensing sequence of a positive measure with 100 point masses and with respect to a random metric. In the 100 point masses, we deal with a mass one in each point of the space and for $\varepsilon = 0.182508$, we plot six condensing iterations: the initial mass, iteration 10, 20, 50, 70 and 102. The limit mass (iter. 102) is an ε-isolated positive measure.

The figures 1.17 present a condensing sequence of a positive measure with respect to a random metric (different as the simulation above). In the 100 point masses, we deal with a mass one in each point of the space and for $\varepsilon = 0.200761$ we plot

(a) First simulations with respect to a random metric:

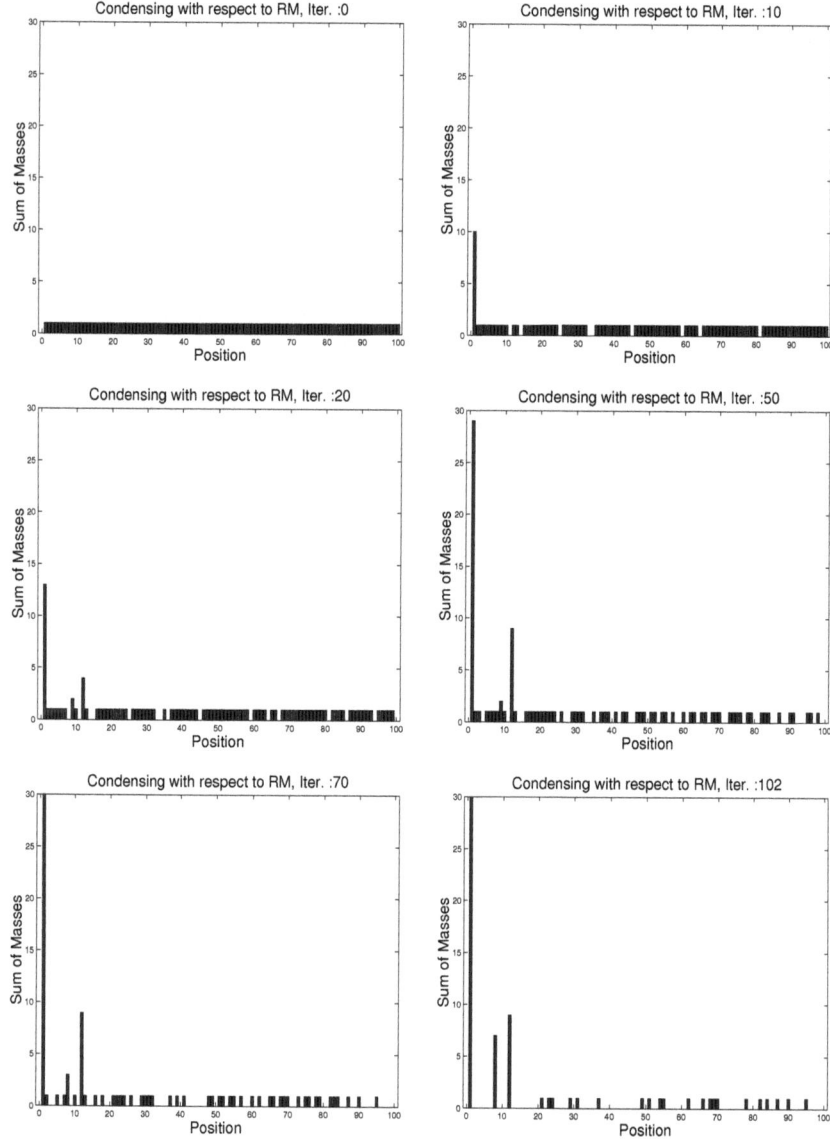

Figure 1.16: *Condensing in a (random) metric space, for $\varepsilon = 0.182508$ (sim. (a))*

1.4. Numerical simulations of condensing sequences

(b) Second simulations with respect to a random metric:

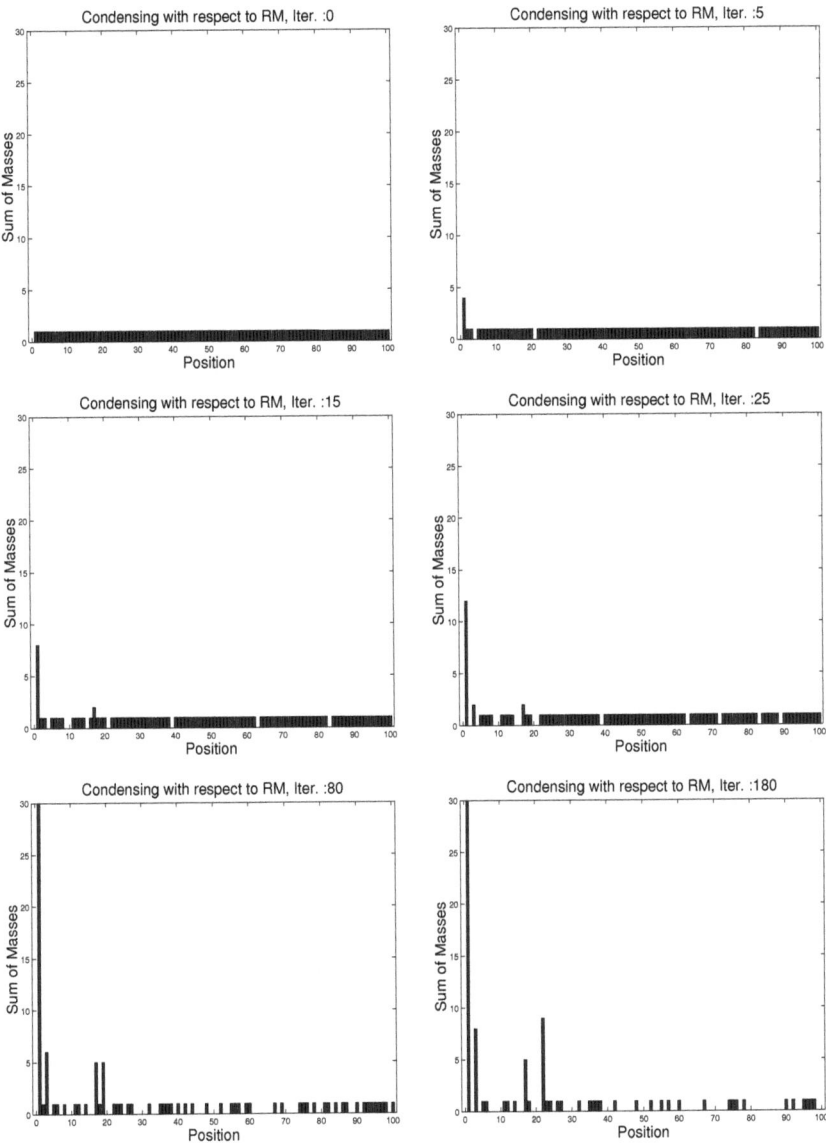

Figure 1.17: *Condensing in a (random) metric space, for $\varepsilon = 0.200761$ (sim. (b))*

six condensing iterations: the initial mass, iteration 5, 15, 25, 80 and 180. The limit mass (iter. 180) is an ε-isolated positive measure.

The figures 1.18 present a condensing sequence of a positive measure with respect to a random metric. In the 100 point masses, we deal with a mass one in each point of the space and for $\varepsilon = 0.254162$ we plot six condensing iterations: the initial mass, iteration 20, 40, 60, 100 and 136. The limit mass (iter. 136) is a consensus.

1.4. Numerical simulations of condensing sequences

(c) Third simulations with respect to a random metric:

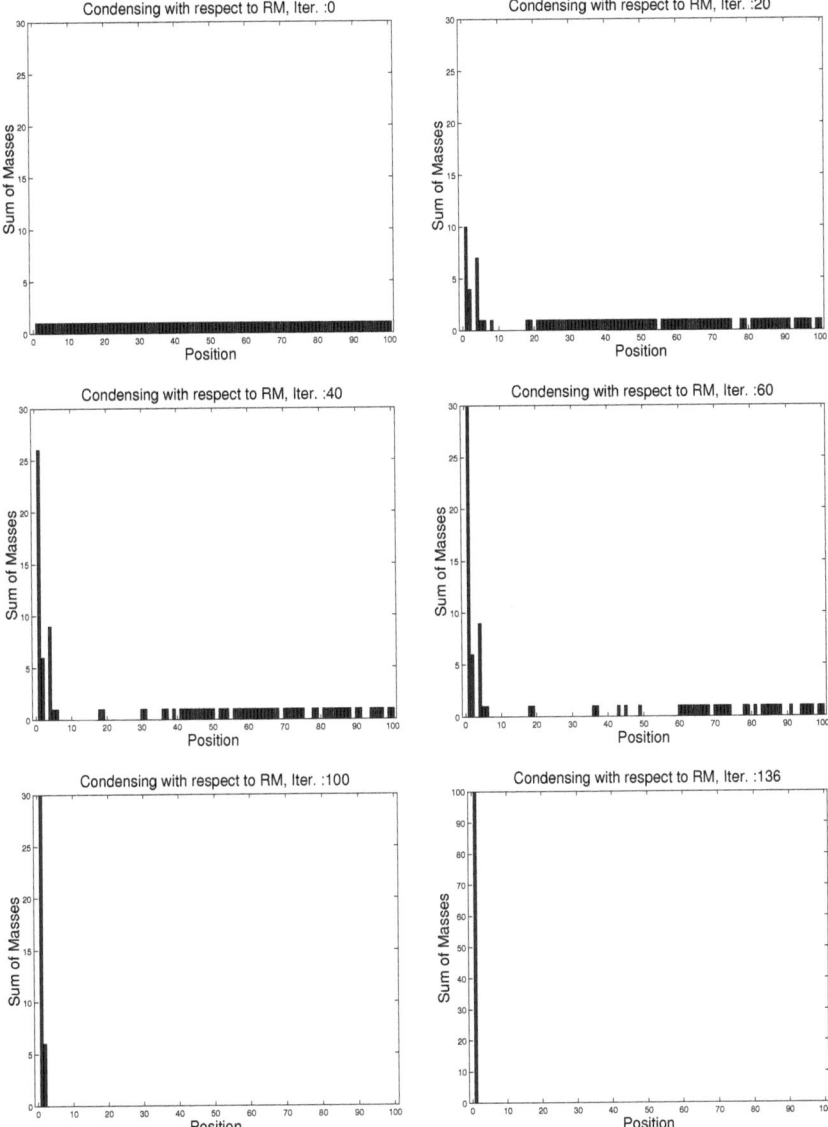

Figure 1.18: *Condensing in a (random) metric space, for $\varepsilon = 0.254162$ (sim. (c))*

1.4. Numerical simulations of condensing sequences

The following plot shows the distance between the particles at the limit state of simulation (a) and (b):

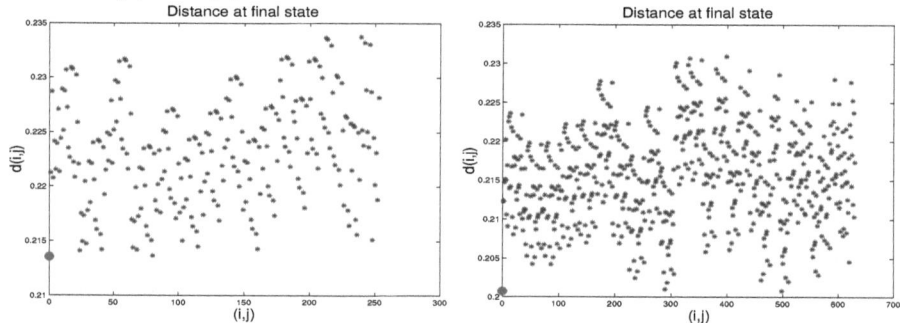

Figure 1.19: *Distances at limit for sim.(a) and (b)*

These plots show the distance check test: Each point is a distance between to particles at the limit (y-axis give it value), the red dot represent the used ε. At the limit all distance are larger than ε.

The following plot shows the decreasing number of strictly positive point masses at each condensing iteration for simulation (a) and (b):

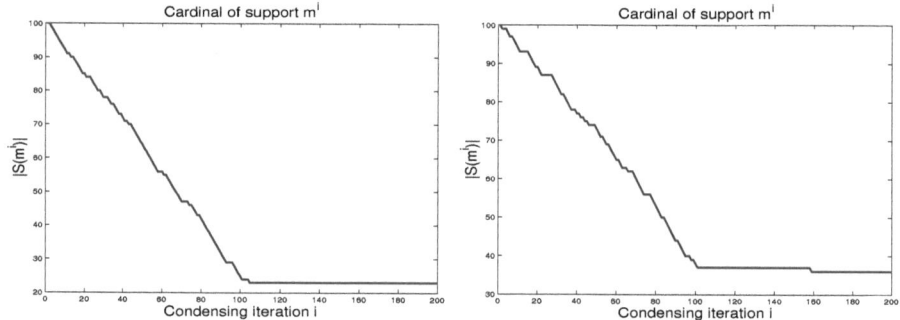

Figure 1.20: *Cardinal of support m^i at limit for sim.(a) and (b)*

The left plot shows the decreasing cardinally of the measures m^i in each condensing iteration.

1.4. Numerical simulations of condensing sequences

The following plot shows the decreasing energy of the simulations (a), (b) and (c):

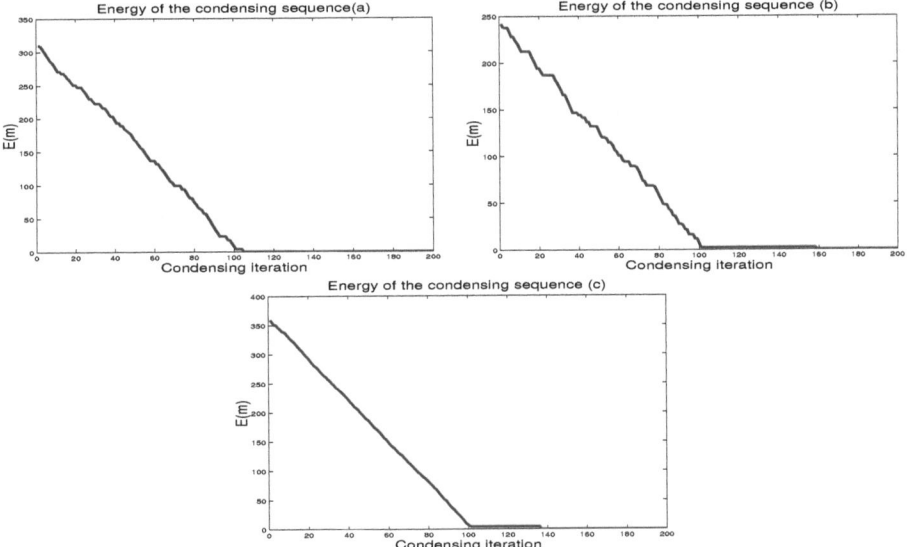

Figure 1.21: *Energy of sim. (a), (b) and (c) in diff. (random) metric spaces.*

The figures 1.21 present respectively the decreasing energy for the three simulations with respect to different random metrics and different values of epsilon. The number of iterations until the limit is differs.

1.4.4 Simultaneously condensing with respect to a random metric

Our goal here is to simulate a large number of simultaneously condensing[2] with respect to a random metric. Therefore, we run our code 1000 times for different random metrics and for the same initial data of a positive measure. Figure 1.22 presents four simulations of them and figure 1.4.4 shows the mean energy of the 1000 simulations.

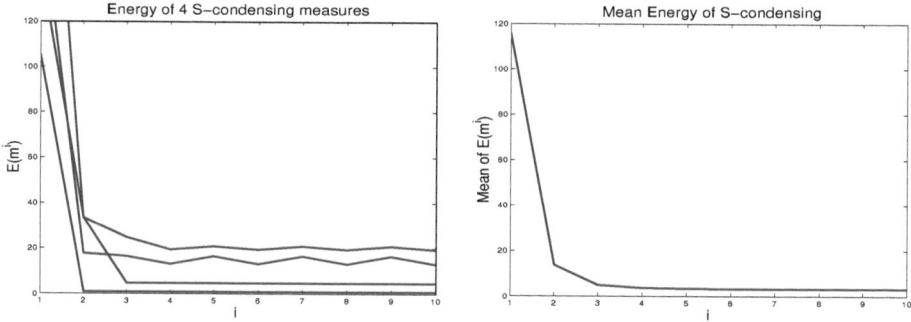

Figure 1.22: *Simultaneously condensing with resp. 1000 RM, Mean energy (right).*

The figure 1.22 shows four curves of energies of four simultaneously condensing of the same measure with uniformly 100 point masses, uniformly distributed. Two curves are decreasing and the others are periodic up to a determined iteration step. Therefore, they are not convergent.

For a measure m with 100 uniform randomly distributed positive discrete measure. The right figure 1.22 presents the mean energy of 1000 simulation of simultaneously condensing m with respect to 1000 different random metrics. It converges to a positive constant. We have remarked that by repeating the simultaneously condensing with respect to random metrics that, the non converging sequence has a periodic energy curve as in figure 1.22.

---
[2] Such as simulations are already presented from Lorenz [19], in Euclidean metric space.

CHAPTER 2

Condensing in continuous metric space

In this chapter we extend the idea of condensing sequences proposed in the previous chapter to a continuous metric space. In order to construct a continuous energy function in m, we introduce the so called "intensity function". Some convergence results for condensing are proved and simulated in many continuous metric spaces.

2.1 Condensing

Let (X, d) be a metric space with metric d. We shall assume that all bounded subsets of X are compact. Hence, X is locally compact and we may use Radon measures. Let $M_+(X)$ be the set of nonnegative Radon measures on X. We shall however for simplicity deal with discrete measures only, given as

$$m := \sum_{x \in S(m)} m(x) \delta_x, \qquad (2.1)$$

where $S(m)$ denote the support of m and δ_x the Kronecker symbol. Note that $S(m)$ is discrete. For such a measure, we define the map E as

$$E : M_+(X) \longrightarrow \mathbb{R}^+ \qquad (2.2)$$

$$E(m) = \sum_{d(x,y) \leq \varepsilon} m(x) m(y) d^2(x, y).$$

In general E is a not continuous function of m.

2.1. Condensing

Example 2.1.1. For $X = \mathbb{R}$, $S(m) = \{1, 2\}$, $\varepsilon = 1$ and a given m with

$$m = \sum_{x \in \{1,2\}} m(x)\delta_x = \delta_1 + \delta_2, \tag{2.3}$$

it follows that $E(m) = 2$. Now let m_j be a sequence of positive measures defined as

$$m_j = \delta_1 + \delta_{2+\frac{1}{j}}, \tag{2.4}$$

it follows

$$\lim_j m_j = m \quad \text{and} \quad E(m_j) = 0, \tag{2.5}$$

and

$$E(\lim_j m_j) = 2 \neq \lim_j E(m_j) = 0. \tag{2.6}$$

Hence, from (2.6), it follows that the map E with the definition (2.2) is not continuous in m.

In order to obtain an energy function which depends continuously on m, we extend the definition (2.2) to the following:

$$E : M_+(X) \longrightarrow \mathbb{R}^+ \tag{2.7}$$

$$E(m) = \int_X \int_X \varphi(x,y) d^2(x,y) m(dx) m(dy)$$

$$= \sum_{x,y} m(x) m(y) \varphi(x,y) d^2(x,y),$$

where φ is a continuous function, which satisfies:

$$\varphi : X \times X \longrightarrow [0,1]; \tag{2.8}$$

$$\varphi(x,y) := \begin{cases} 1, & \text{if } d(x,y) \leq \varepsilon, \\ 0, & \text{if } d(x,y) \geq \varepsilon + \theta. \end{cases}$$

For $\varepsilon > 0$ and $\theta > 0$.

Remark 2.1.1. The parameters $\varepsilon > 0$ and $\theta > 0$ will be fixed throughout this chapter. The function φ will be called **intensity function**.

2.1. Condensing

Example 2.1.2.

1. For $\varepsilon, \theta > 0$, define the function ψ as

$$\psi : [0, \infty) \longrightarrow [0, 1];$$

$$\psi(x) := \begin{cases} 1, & \text{if } 0 \le x \le \varepsilon, \\ 0, & \text{if } x \ge \varepsilon + \theta. \end{cases}$$

Then $\varphi : (x, y) \mapsto \varphi(x, y) = \psi(d(x, y))$ is an intensity function.

2. The following function is an intensity function for every $\varepsilon > 0$ and $\theta > 0$:

$$\varphi : \mathbb{R} \times \mathbb{R} \longrightarrow [0, 1];$$

$$\varphi(x, y) := \begin{cases} 1, & \text{if } |x - y| \le \varepsilon, \\ 0, & \text{if } |x - y| \ge \varepsilon + \theta. \\ \left(\frac{\varepsilon + \theta - |x - y|}{\theta}\right) & \text{otherwise.} \end{cases}$$

Let $M_E(X)$ be the set of discrete and nonnegative measure with $E(m) < \infty$. As in the previous chapter a pair $(a, a^*) \in X \times X$ operates on $M_E(X)$ as

$$m^*(x) = (a, a^*, m)(x) = \begin{cases} m(x); & \text{if } x \notin \{a, a^*\}, \\ 0; & \text{if } x = a, \\ m(a) + m(a^*); & \text{if } x = a^*. \end{cases}$$

Note that if $m \in M_E(X)$ then $m^* \in M_E(X)$.

Definition 2.1.1. The energy of a point $a \in X$ with respect to $m \in M_E(X)$ is

$$e : X \times M_E(X) \longrightarrow \mathbb{R}^+ \tag{2.9}$$

$$e(a, m) = \sum_y m(y) \varphi(a, y) d^2(a, y).$$

2.1. Condensing

Lemma 2.1.1. For $a, a^* \in X$, $m \in M_E(X)$ and $m^* := (a, a^*, m)$, we have

$$E(m) - E(m^*) = 2m(a)\big[e(a, m) - e(a^*, m^*)\big]. \qquad (2.10)$$

Proof. For simplicity, let us denote by I_m the following term:

$$I_m := \sum_{\{x,y\} \cap \{a, a^*\} = \emptyset} m(x)m(y)\varphi(x,y)d^2(x,y),$$

and let us compute the energy of m:

$$\begin{aligned}
E(m) &= \sum_{x,y} m(x)m(y)\varphi(x,y)d^2(x,y) \qquad (2.11)\\
&= I_m + 2\sum_{x,y} m(a)m(y)\varphi(a,y)d^2(a,y)\\
&\quad + 2\sum_{x,y} m(a^*)m(y)\varphi(a^*,y)d^2(a^*,y) - 2m(a^*)m(a)\varphi(a^*,a)d^2(a^*,a)\\
&= I_m + 2m(a)e(a,m) + 2m(a^*)e(a^*,m) - 2m(a^*)m(a)\varphi(a^*,a)d^2(a^*,a).
\end{aligned}$$

Similarly for m^* (by replacing m by m^*), we get

$$\begin{aligned}
E(m^*) &= I_{m^*} + 2m^*(a)e(a, m^*) + 2m(a^*)e(a^*, m^*) \qquad (2.12)\\
&\quad - 2m^*(a^*)m^*(a)\varphi(a^*,a)d^2(a^*,a).
\end{aligned}$$

Note that

$$m^*(a) = 0 \quad \text{and} \quad I_m = I_{m^*}, \qquad (2.13)$$

and in addition we have:

$$e(a^*, m^*) = e(a^*, m) - m(a)\varphi(a^*,a)d^2(a^*,a). \qquad (2.14)$$

Therefore, from (2.12)–(2.14) it follows:

$$E(m^*) = I_{m^*} + 2m^*(a)e(a, m^*). \qquad (2.15)$$

From (2.11) and (2.15), it follows that

$$E(m) - E(m^*) = 2m(a)\big[e(a,m) - e(a^*,m^*)\big], \qquad (2.16)$$

which prove the result of the lemma.

2.1. Condensing

□

In order to extend the move notion already defined in the previous chapter to a continuous metric space, we use the intensity function to introduce an additional necessary condition. Recall that in the finite case for (m, m^*) to be condensing, we required

(i) $e(a, m) - e(a^*, m^*) > 0$ and

(ii) $d(a, a^*) \leq \varepsilon$.

Now we require a^* minimizing $y \to e(y, m^*) \varphi(a, y)$.

Definition 2.1.2. *A pair (m, m^*) is called condensing, if there is $(a, a^*) \in X \times X$ such that*

(i) $m^* = (a, a^*, m)$ *and*

(ii) a^* *minimizing* $y \to e(y, m^*) \varphi(a, y)$, *with* $d(a, y) \leq \varepsilon + \theta$.

A sequence of non negative measures $m^1, m^2, \ldots$ is called condensing, if for every i the pair (m^i, m^{i+1}) is condensing. A mapping f such that $m^* = f(m)$ is called condensing.

Remark 2.1.2. Let (X, d) be a metric space. $m^1, m^2, \ldots$ is said to be condensing, if for every i there is a mapping f_i, which is condensing and such that

$$m^{i+1} = f_i(m^i). \tag{2.17}$$

A condensing mapping f is said to be **singularly** condensing if $f(y) \neq y$ for only one $y \in S(m)$. A condensing sequence is called singular if every f_i is singular. f is said **simultaneously** condensing, if the conditions of definition (2.1.2) are

2.1. Condensing

required for all $x \in S(m)$. Note that in each condensing step, we define the image measure operated by a singularly condensing mapping as

$$f : M_+(X) \longrightarrow M_+(X); \qquad (2.18)$$

$$f(m)(x) := \begin{cases} m(f(x)), & \text{if } f(x) \notin S(m), \\ m(f(x)) + m(z), & \text{if } f(x) = z \in S(m). \end{cases}$$

The total energy is given by

$$E_\varepsilon(f(m)) = \sum_{x,y} m(f(x))m(f(y))\varphi(f(x),f(y))d^2(f(x),f(y)). \qquad (2.19)$$

Remark 2.1.3. *The energy (2.19) can be written in integral form as*

$$E(f(m)) = \int_X \int_X \phi(f^2(x), f^2(y))m(dx)m(dy), \qquad (2.20)$$

where $\phi(x,y) := d^2(x,y)\varphi(x,y)$ and $f^2 = f \circ f$.

Proof. By definition of the total energy, we have

$$E(f(m)) = \int_X \int_X d^2(f(x),f(y))\varphi(x,y)d^2(f(x),f(y))f(m)(dx)f(m)(dy).$$

Then

$$\begin{aligned}
E(f(m)) &= \int_X \int_X \phi(f(x),f(y))f(m)(dx)f(m)(dy) \\
&= \int_X \left(\int_X \phi(f(x),f(y))f(m)(dy) \right) f(m)(dx) \\
&= \int_X \left(\int_X \phi(f(x),f^2(y))m(dy) \right) f(m)(dx) \\
&= \int_X \left(\int_X \phi(f(x),f^2(y))f(m)(dx) \right) m(dy) \\
&= \int_X \int_X \phi(f^2(x),f^2(y))m(dx)m(dy).
\end{aligned}$$

$\square$

2.2 Convergence

In this section we prove some convergence results of condensing sequences on continuous metric space.

Definition 2.2.1. Let $(m^i)_{i\geq 1}$ be a condensing sequence. (m^i) is called

(i) *Consensus*, if $\sharp S(\lim_i m^i) = 1$.

(ii) *Segregation*, if $\sharp S(m^i) = \sharp S(m^1)$, $\forall i$. Note that a condensing sequence with a no collision condition leads to a segregation.

where $\lim_i m^i$ is the limit measure of the condensing sequence, since $\sharp S(m^i)$ is bounded (discrete), $\sharp S(\lim_i m^i)$ is also bounded.

Corollary 2.2.1. If $m^1, m^2, \ldots$ is a singularly condensing sequence, then

$$\lim_i E(m^i) = \ell \geq 0. \tag{2.21}$$

Proof. The proof follows immediately from lemma 2.1.1.

□

In order to prove the convergence of a singular condensing, we define the so called effectively condensing sequences:

Definition 2.2.2. A singularly condensing sequence $m^1, m^2, \ldots$ is called effectively condensing sequence if there exists $c > 0$ such that

(i) $m^{i+1} = (a, a^*, m^i)$,

(ii) $E(m^i) - E(m^{i+1}) \geq c\alpha(m^i)$,

where

$$\alpha(m^i) = \max_y \{\varphi(y, a) d^2(y, a) | y \in S(m^i)\}.$$

2.2. Convergence

Remark 2.2.1. From (2.15) and (2.16) and

$$c = 2\min_a \{m(a) | a \in S(m^i)\},$$

follow the existence of the effectively condensing sequence.

Lemma 2.2.1. *Suppose that $m^1, m^2, \ldots$ is an effectively condensing.*

$$\text{if } \lim_{i \to \infty} E(m^i) = E(m), \text{ then } \alpha(m) = 0. \tag{2.22}$$

Proof. Consider $\eta > 0$, from

$$\lim_{i \to \infty} E(m^i) = E(m),$$

exists i_0 such that for all $i > i_0$, it follows

$$2c\alpha(m^i) \leq E(m^i) - E(m) \geq \eta, \tag{2.23}$$

for large i, we get $\lim_i \alpha(m^i) = 0$ and still $\alpha(m) = 0$.

$\square$

Theorem 2.2.1. *If (X, d) is a compact metric space, then every effectively condensing sequence $m^1, m^2, \ldots$ is finite such that*

$$\{m^1, m^2, \ldots\} = \{m^1, m^2, \ldots, m^k\} \quad \text{and} \quad E(m^k) = 0. \tag{2.24}$$

Proof. Since X is compact, then

$$\cup_i S(m^i)$$

is relative compact and there exists a subsequence m^{i_j} of m^i such that

$$\lim_j m^{i_j} = m \quad \text{and} \quad \lim_j E(m^{i_j}) = E(m).$$

From lemma 2.2.1 it follows that

$$\lim_j \alpha(m^{i_j}) = 0 \quad \text{and} \quad \alpha(m) = 0.$$

Since the sequence $m^1, m^2, \ldots$ is effectively condensing, and from definition 2.2.2 there exists k such that $\alpha(m^k) = 0$. Therefore, for all $x, y \in S(m^k)$, it follows

$$d(x,y) = 0 \quad \text{or} \quad d(x,y) \geq \varepsilon + \theta. \tag{2.25}$$

Hence, $E(m^k) = 0$ and still m^k is a collection of isolated masses with propriety (2.25) or a point mass $m = m(X)\delta_a$ for $a \in X$.

□

2.3 Numerical Simulations

This section presents numerical simulations of condensing sequences on a continuous metric space. To give sense to the condensing sequence (2.1.2) two steps are required namely, the computation of local energy given in (2.9) of a particle x and the comparison procedure with the local energy of a randomly chosen point mass y such that $d(x,y) \leq \varepsilon + \theta$. It is important to note that the particle y minimizes the energy on his neighborhood. The energy in the H.K. model given in [15] will be minimized on the neighborhood of the particle x. For more details see the works of [15, 16, 17, 19]. In our model, the energy is minimized in y, if x moves to y, where the positions are randomly chosen, which satisfied the conditions of an admissible move. In general we use the uniform and normal distributions on the $\varepsilon-$neighborhood of the particle and for our random generator, we use the construction mentioned in my diploma thesis[25]. We simulate the condensing sequences in three domains: In the real line, the unit circle considered as a one dimensional manifold and in the real plane. In order to give an experimental sense for simultaneously condensing, we simulate the H.K. model in a one dimensional manifold. The summary of the numerical simulations is listed as follow:

2.3. Numerical Simulations

1. **Simulations on the real line:**

    (a) Uniform mass distribution.

    (b) Uniform random mass distribution.

2. **Simulations on unit circle (one dimensional manifold):**

    (a) Uniform mass distribution with mass one.

    (b) Uniform random mass distribution.

3. **Simulations on the real plane:**

    (a) Uniform random mass distribution.

    (b) Uniform random mass distribution.

4. **Simulation of segregation sequences.**

    (a) Uniform random mass distribution.

    (b) Uniform mass distribution.

5. **simultaneously condensing on the unit circle.**

2.3.1 Condensing on the real line

This section presents simulations on the real line. In order to compute the density of the points masses of a measure at each iteration, the space domain is the discretized into Nx uniform gridpoints. We carry out two tests of condensing sequences of two measures given as:

$$m := \sum_{x \in S(m) \subset [0,1]} m(x) \delta_x, \qquad (2.26)$$

2.3. Numerical Simulations

where the metric space (X, d) and m have the proprieties:

$$X = [0, 1], \qquad (2.27)$$
$$d(x, y) = |x_i - x_j|, \forall x, y \in \mathbb{R}.$$
$$S(m) = \{x_1, \ldots, x_{50}\},$$
$$Nx = 50.$$

Since $[0, 1] \subset \mathbb{R}$, we use the absolute value as a metric in X. The emergence of $S(m)$ requires the extension of the computation domain, namely even if the initial spatial positions are chosen in $[0, 1]$, the final measure has not necessary positions in $[0, 1]$ only. We run our code after fixing the order of reactions (the array of 50 index will randomly permuted). The following table summarizes the results of the simulations on the real line:

Parameter/Sim.	(a)	(b)
ε	0.02	0.02
θ	0.01	0.01
Initial state	50 masses (one)	50 masses ($U(0, 4)$)
Final state	4 isolated masses	4 isolated masses
Time in Sec	1019	1158
Number of iterations	176	196

Table 2.1: Results of simulations in the real line.

The figures 2.1 present the density of a measure with 50 point masses in twelve condensing iterations (iter. 0, 5, 10, 20, 25, 40, 30, 40, 50, 90, 100 and 175), for $\varepsilon = 0.02, \theta = 0.01$, the first and the last figures present resp. the initial and the limit measure. The initial state is a deterministic distribution of masses and the limit measure (inter. 175) is constituted only from $\varepsilon + \theta$ isolated masses. The figures 2.2 present the density of measure with 50 point masses in twelve

(a) Condensing of a measure with uniform initial distribution of masses:

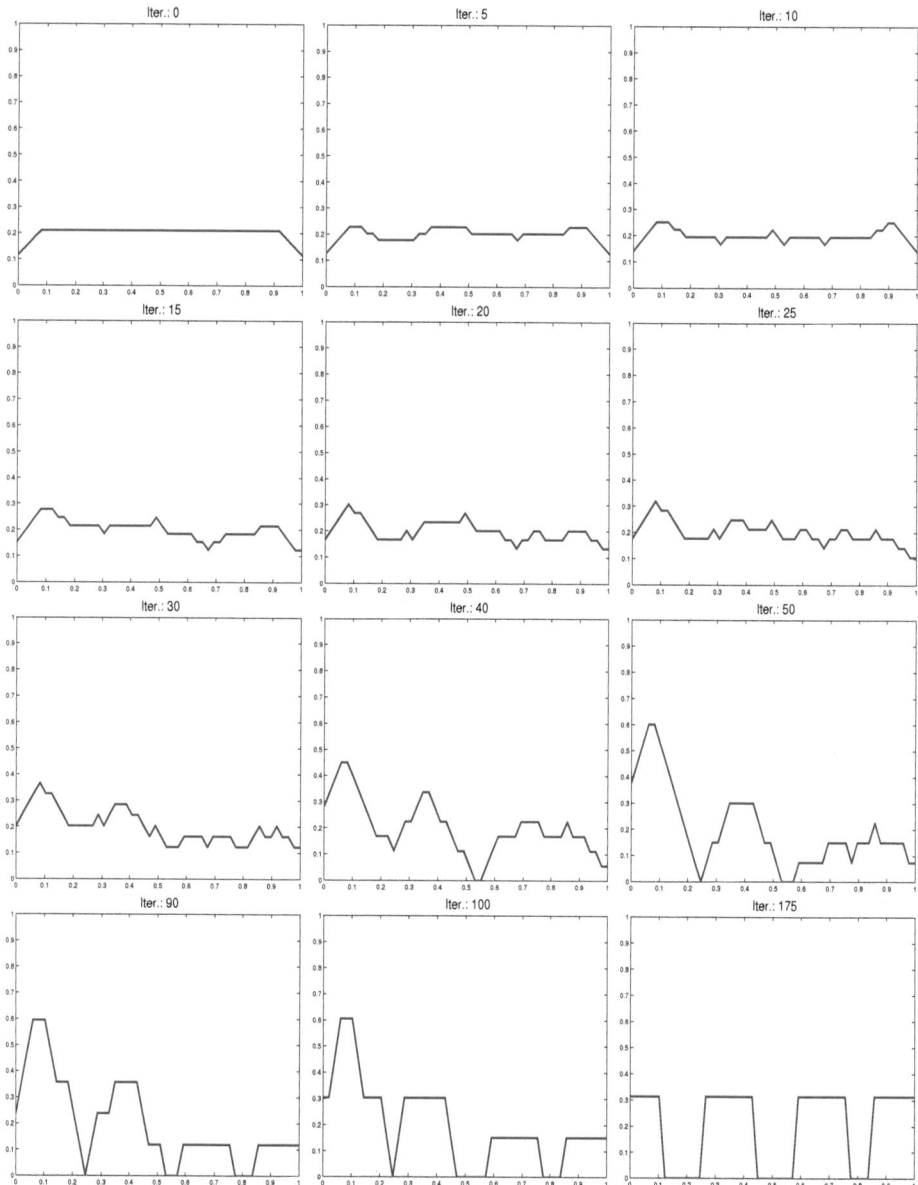

Figure 2.1: *Density of condensing measures on the real line (sim. (a))*.

2.3. Numerical Simulations

(b) Condensing of a measure with random initial distribution of masses:

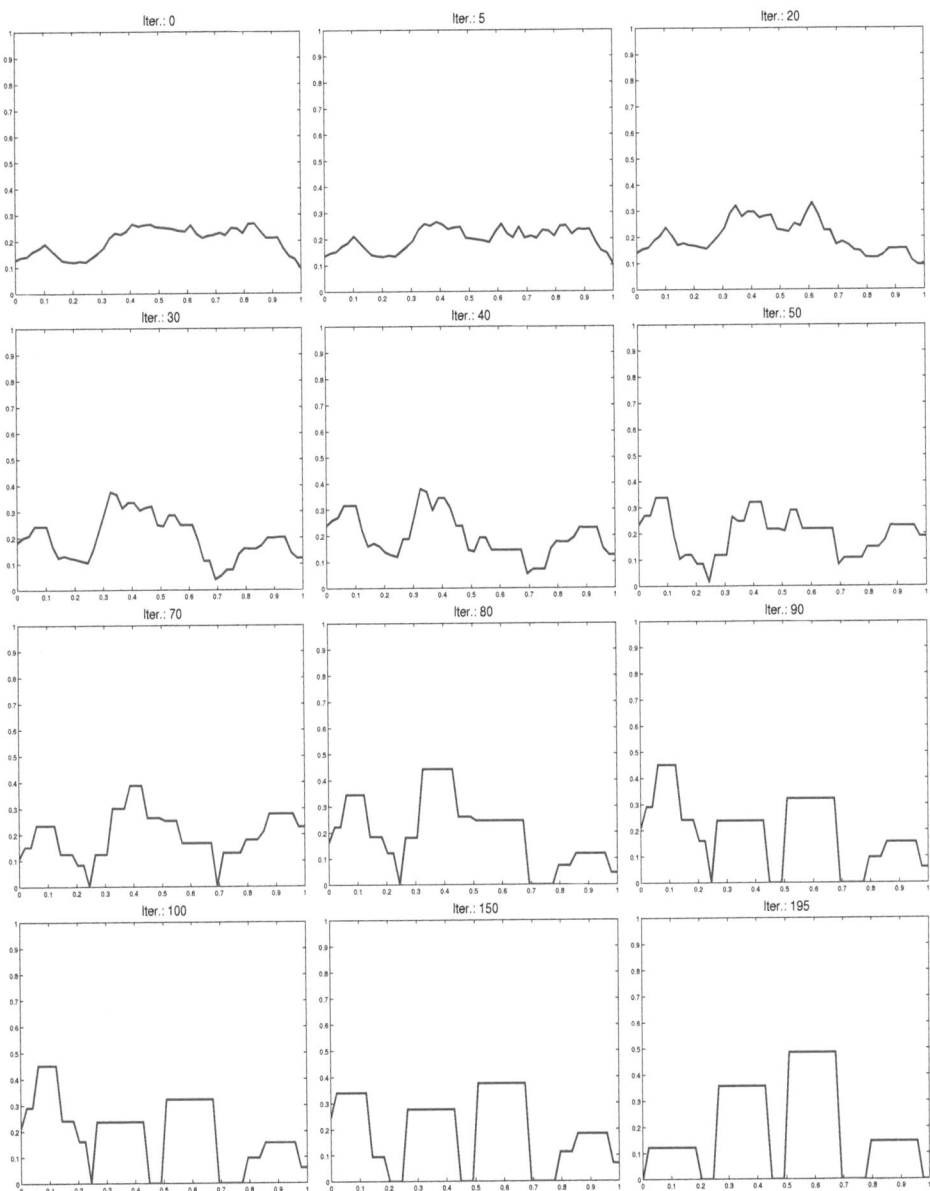

Figure 2.2: *Density of condensing measures on the real line (sim. (b))*

2.3. Numerical Simulations

condensing iterations (iter. 5, 20, 30, 40, 50, 70, 80, 90, 100, 150 and 195) in the real line, for $\varepsilon = 0.02, \theta = 0.01$. The first and the last figures present resp. the initial and the limit measures. The initial state is generated with a uniform random distribution ($U(0,4)$ (on the figures, the density is normalized) and the final state (iter. 195) is constituted only from $\varepsilon + \theta$ isolated masses.

Remark on the results 2.3.1. The figures 2.1 and 2.2 have different initial mass but they have the same condensing behavior. The Final states are isolated point masses with different supports and different masses.

2.3.2 Condensing on the unit circle

In this section, we run our code on the unit circle for a measure given by

$$m := \sum_{x \in S(m) \subset S^1} m(x)\delta_x, \qquad (2.28)$$

where the metric space $X = (S^1, d)$ is given such that:

$$d(x,y) = g(x,y), \quad \forall x, y \in S^1, \qquad (2.29)$$

$$S(m) = \{x_1, \ldots, x_{100}\}, \quad Nx = 50.$$

The metric used here is a geodesic[1] metric, given by

$$g(x,y) = \cos^{-1}(<x,y>) \in [0, \pi],$$

where $<\cdot,\cdot>$ is the usual scalar product in $\mathbb{R}^2$. For NP= 50, $(\varepsilon, \theta) = (\frac{\pi}{4}, 0.001)$ resp. $(\varepsilon, \theta) = (\frac{\pi}{8}, 0.001)$, the figures 2.3 resp. 2.4 present simulations of condensing of particles on the unit circle. The figures show six time iterations the spatial positions of the particles. Our code stops if the total energy achieved some given negligible tolerance. The following table summarizes the result of the results i S^1:

Parameter/Sim.	(a)	(b)
ε	$\frac{\pi}{16}$	$\frac{\pi}{8}$
θ	0.001	0.001
Initial state	50 masses (one)	50 masses $(U(0,4))$
Final state	4 isolated masses	12 isolated masses
Time in Sec	910	851
Number of iterations	4900	3950

Table 2.2: Results of simulations in S^1.

---

[1]The shortest path between two points.

Remark on the results 2.3.2. Because of the construction of the unit circle (compact and bounded metric space), and for a large number of particles, the Final state of the condensing sequence of particles is at the limit an isolated mass points. These are clearly shown by figures 2.3 and 2.4. We also observe that the the cardinal of the support desponds on the choice of ε. In fact that $\varepsilon_1 = \frac{\pi}{16} < \varepsilon_2 = \frac{\pi}{8}$, the finale state of the first simulation has a larger support as the second one.

The figures 2.3 present six condensing iterations on the unit circle considered as one dimensional manifold. We show the spatial positions of the iterations 50, 100, 500, 1000 and 4900. The limit mass (iter. 4900) is constituted of many $\varepsilon + \theta$ isolated groups of point masses. We have plotted only the center of mass of each subgroup. The figures 2.4 present six condensing iterations on the unit circle considered as one dimensional manifold. We show spatial position of the particles in many iterations, namely 100, 200, 300, 2000 and 3950. The limit mass (iter. 3950) is constituted of eleven $\varepsilon + \theta$ isolated groups of point masses. We have plotted only the center of mass of each subgroup.

2.3. Numerical Simulations

(b) Simulation on one dimensional manifold with deterministic initial distribution:

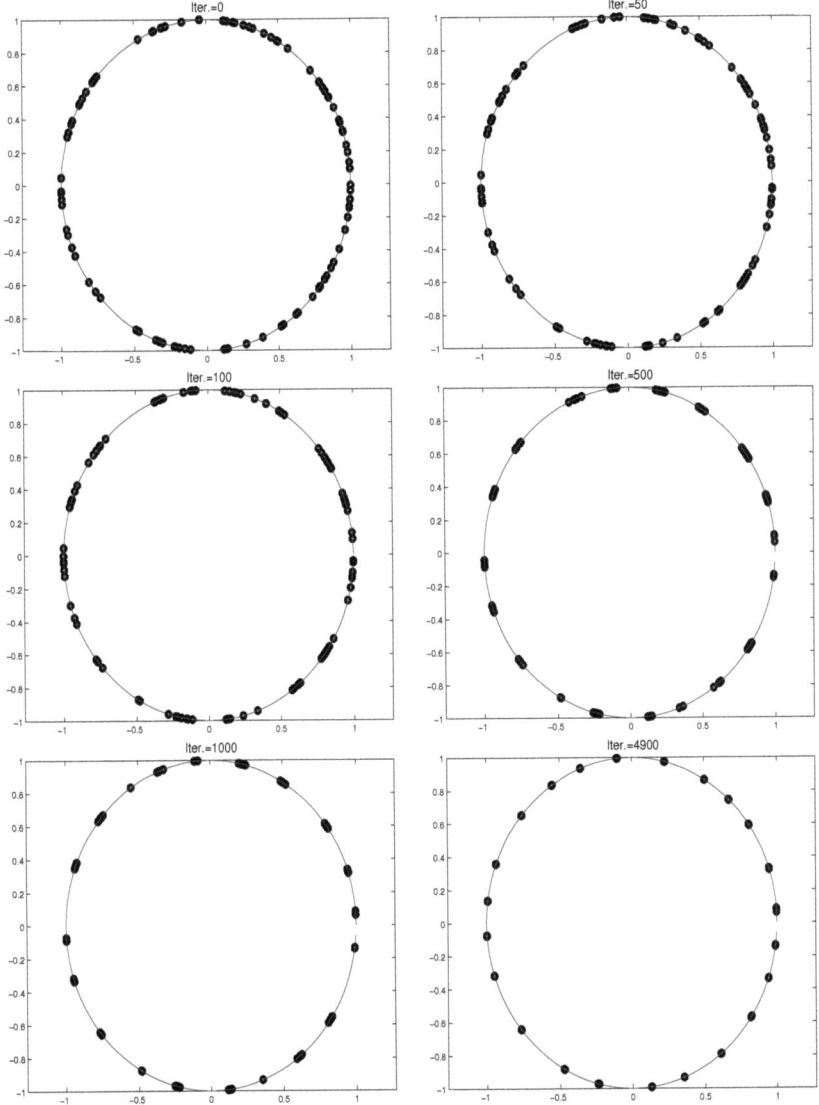

Figure 2.3: *Condensing of particles on a one dimensional manifold (sim. (a)).*

(b) Simulation on a one dimensional manifold with random initial distribution:

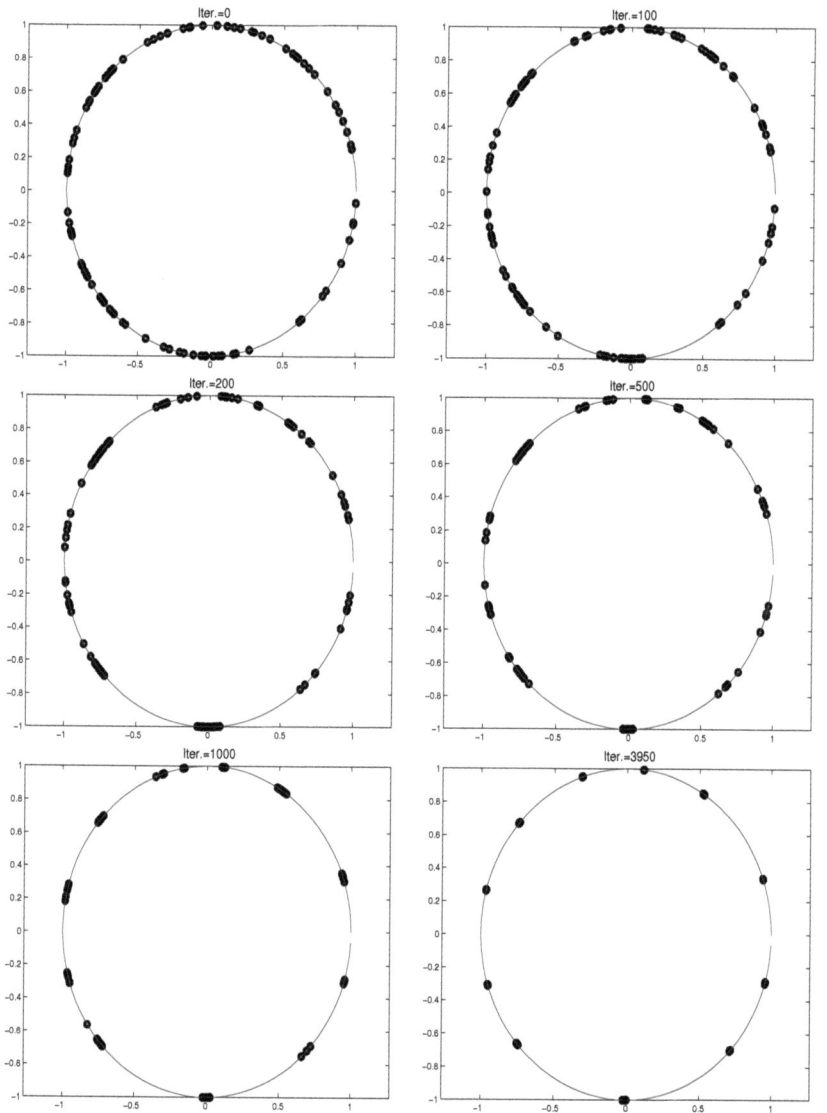

Figure 2.4: *Condensing of particles on a one dimensional manifold (sim. (b))*.

2.3. Numerical Simulations

Simulation on a one dim. manifold (a) and (b):

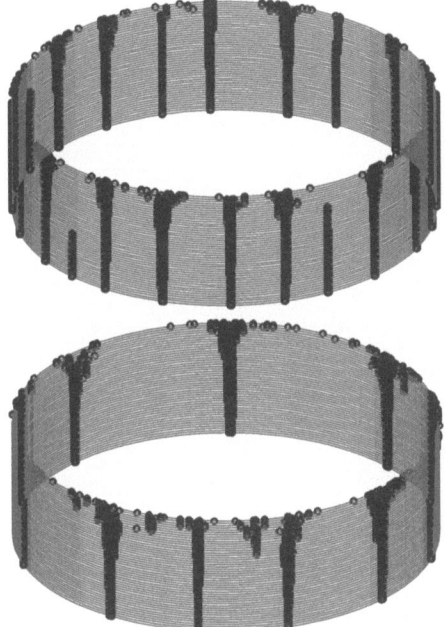

Figure 2.5: *Condensing of particles on a one dim. manifold (sim. (a) and (b))*.

This figures show 100 condensing steps of simulation (a) and (b). One can easily constat the by looking for position with minimal energy a jump moves are possible. These jumps don't appear in the HK-model.

2.3.3 Condensing on the real plane

This section presents simulations on the real plane. In order to compute the density of the points masses of a measure at each iteration, the space domain is the discretized into $Nx \times Nx$ uniform two dimensional gridpoints. We carry out two tests of condensing sequences of two measures given as:

$$m := \sum_{x \in S(m) \subset [0,1]^2} m(x)\delta_x. \tag{2.30}$$

The metric space $X = [0,1]^2$, is constructed as following:

$$d(x,y) = \|x-y\|_2, \quad \forall x, y \in \mathbb{R}^2, \tag{2.31}$$

$$S(m) = \{x_1, \ldots, x_{441}\}, \quad Nx = 20. \tag{2.32}$$

The emergence of $S(m)$ requires the extension of the computational domain, namely even if the initial spatial positions are chosen in $[0,1]^2$, the limit measure has not necessary positions in $[0,1]$. Therefore, the figures are plotted on an extension domain of $[0,1]^2$, namely the domain $[-2,3]^2$. We run our code after fixing the order of reactions (the array of 441 index will be randomly permuted). The following table summarizes the results of the simulations on the real plane:

Parameter/Sim.	(a)	(b)
ε	0.1	0.1
θ	0.001	0.001
Initial state	441 masses (one)	441 masses ($U(0,4)$)
Final state	condensing	condensing
Time in Sec	more than 7102	more than 10760
Number of iterations	more than 9000	more than 20000

Table 2.3: Results of two simulations on the real plane.

2.3. Numerical Simulations

Remark on the results 2.3.3. For $(\varepsilon, \theta) = (0.1, 0.001)$. The figure 2.6 present the two dimensional density of the measure and in order to show the emergence of the particle (here consensus), we plot the corresponding contour in figure 2.7. This simulation needs about 9000 time iterations until the last iteration presented in the last figure. Our code breaks down, when the total energy will be negligible (compared with a fixed tolerance). It is clearly shown in figures 2.6 and 2.8 that the particles build a mass points with hight density in the middle of the computation domain. This result will be different if we simulate another simulation, even if we use the same data for the initial measure.

(a) First simulation on real plane, with stochastic initial distribution:

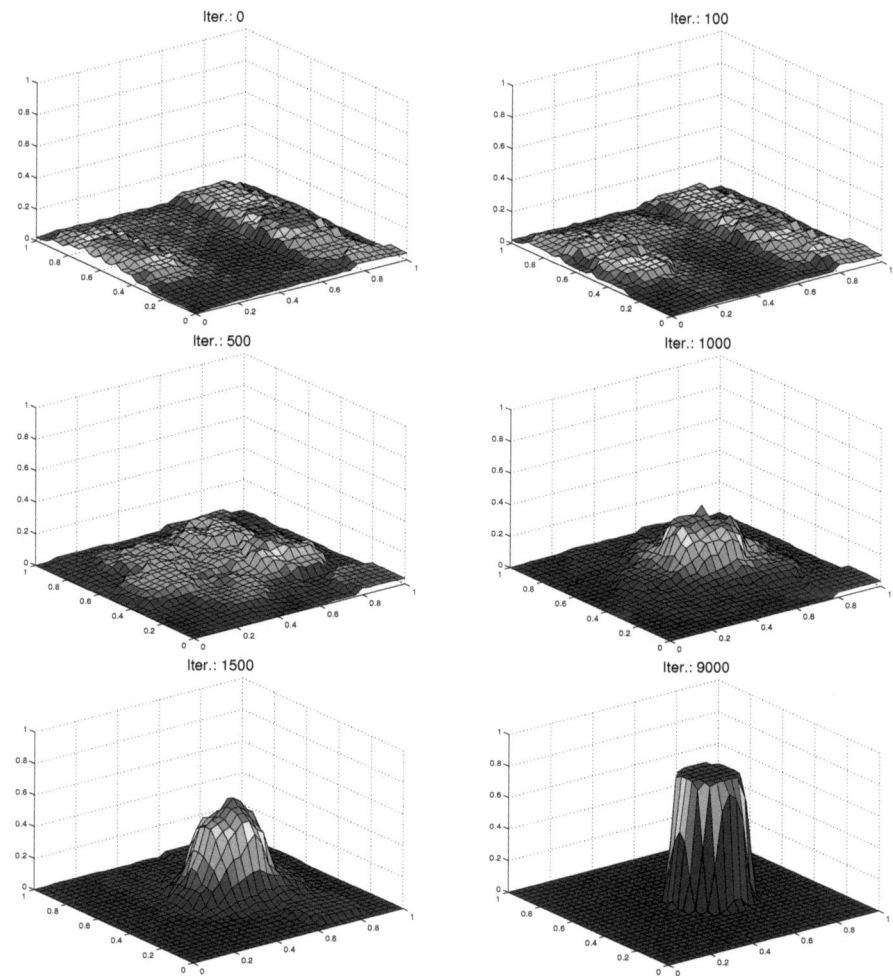

Figure 2.6: *Density of condensing of particles on real plane (sim. (a))*

Thes figures present the density of six condensing iterations 100, 500, 1000, 1500 and 9000. We show the condensing of the density on the middle.

2.3. Numerical Simulations

(a) First simulation on real plane, with stochastic initial distribution:

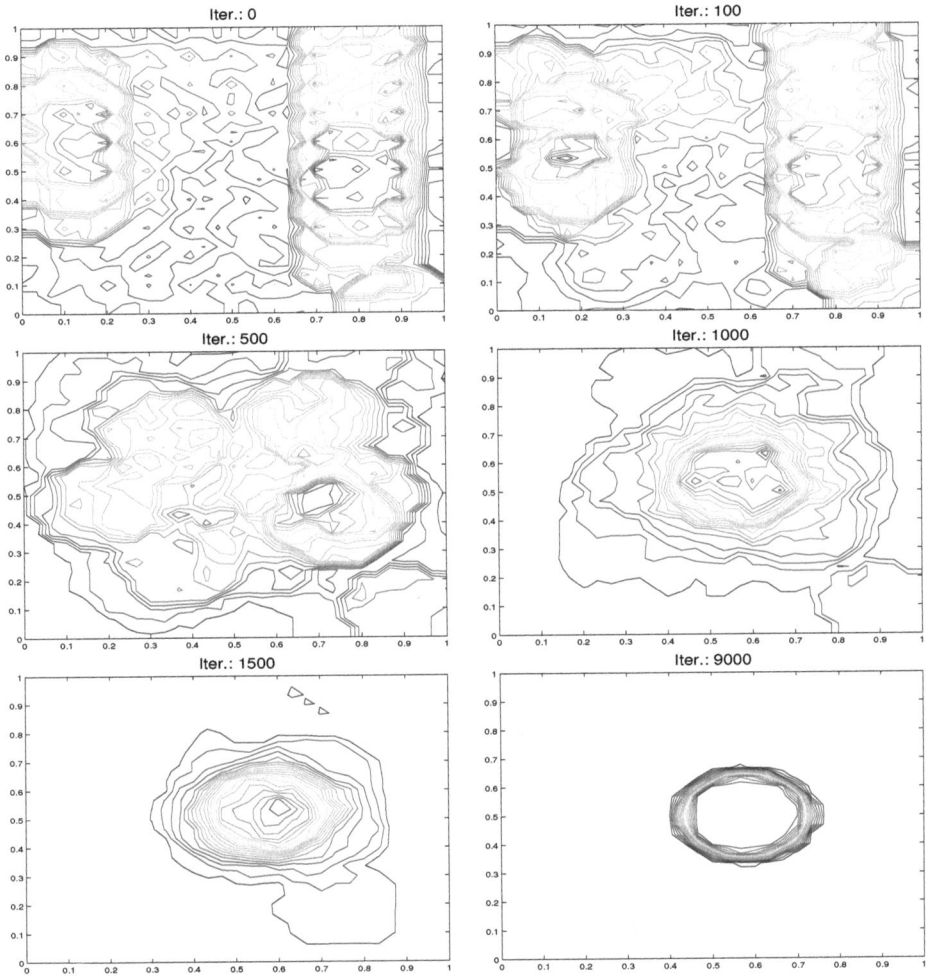

Figure 2.7: *Contour of the density of the condensing on real plane (sim. (a))*

These figure present the contour density of the six condensing iterations (a), we show the initial and the Final mass and also the iterations 100, 1000, 1500 and 9000.

(b) Second simulation on real plane, with stochastic initial distribution:

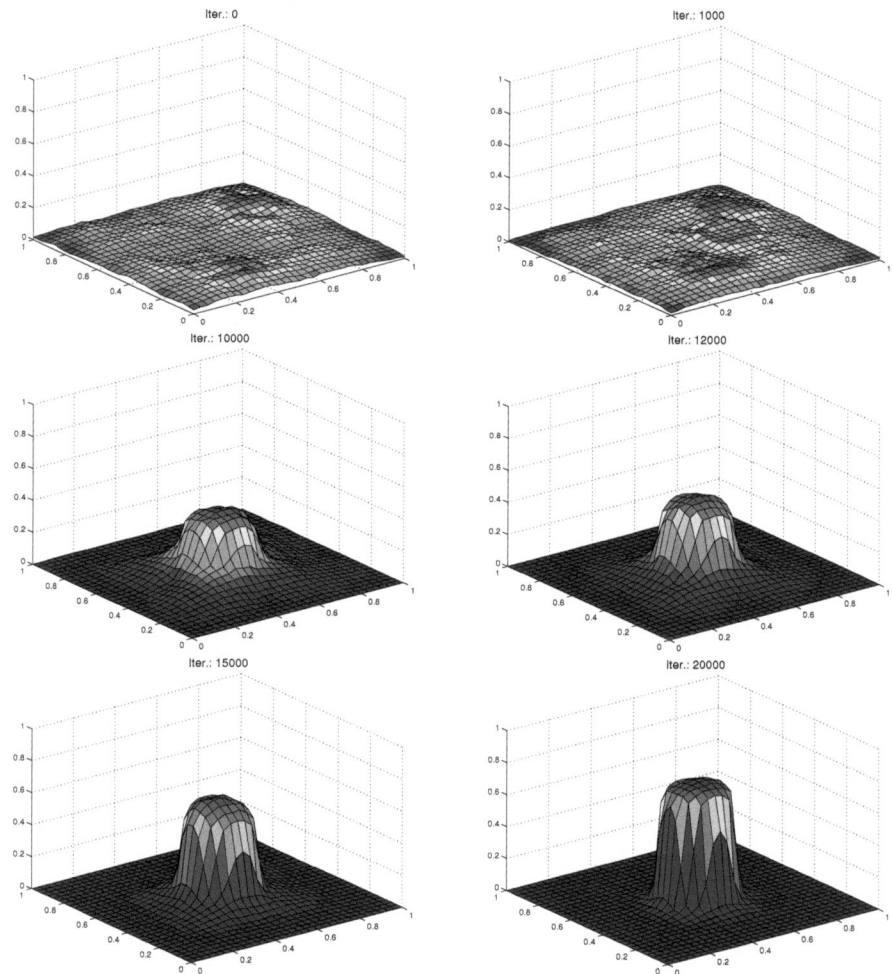

Figure 2.8: *Density of condensing of particles on real plane (sim. (b))*

These figures present the density of six condensing iterations, we show the initial and the Final mass and also the iteration 1000, 10000, 12000, 15000 and 20000. The mass is concentrated on one point on the middle of the computational domain.

2.3. Numerical Simulations

(b) Second simulation on real plane, with stochastic initial distribution:

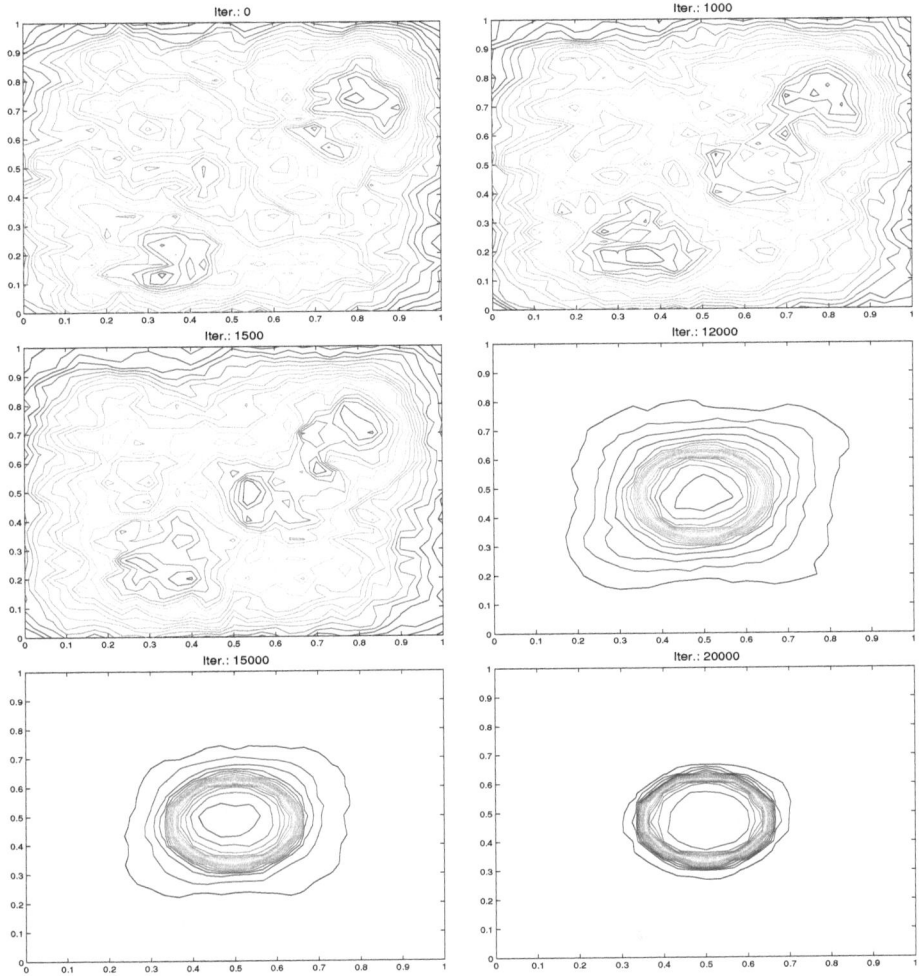

Figure 2.9: *Contour of the density of the condensing on real plane (sim. (b))*

These figures present the contour density of the six condensing iterations (a), we show the initial and the Final mass and also the iterations 1000, 1500, 12000, 15000 and 20000. The final state looks like a translation of mass in the middle of the computational domain.

2.3.4 Segregation sequences

In this section, we construct a segregation sequence. Therefore, we chose NP particles uniformly distributed and for our computational domain, we use an equidistance space discretization of the domain $[0,1]^2$. Each gridpoints has a mass one. We run our code by fixing the non random order of reactions. Each particle changes its position without collision with other particles for a measure

$$m := \sum_{x \in S(m) \subset [0,1]^2} m(x)\delta_x. \quad (2.33)$$

The metric space is constructed as follows:

$$X = [0,1]^2, \quad (2.34)$$
$$d(x,y) = \|x-y\|_2, \quad \forall x,y \in \mathbb{R}^2,$$
$$S(m) = \{x_1, \ldots, x_{400}\}, \quad (2.35)$$
$$Nx = 20.$$

The organizing process of this group is computed in $[-2,3]^2$. The following table summarizes the results of the simulations of segregation sequences:

Parameter/Sim.	(a)	(b)
ε	0.1	0.2
θ	–	–
Initial state	(Random) 400 masses (one)	(Uniform) 400 masses (one)
Final state	segregation	segregation
Time in Sec	more than 7102	more than 10760
Number of iterations	more than 20000	more than 20000

Table 2.4: Results of two simulations in $\mathbb{R}^2$ (segregation).

Remark on the results 2.3.4. In figure 2.10 and 2.11, we plot respectively the geographical positions of all particles and the corresponding two dimensional

2.3. Numerical Simulations

densities. The results obtained by choosing $\varepsilon = 0.1$ resp. $\varepsilon = 0.2$ and NP=400 particles. Different realizations are plotted. In these figures, we also show the first and the last time realizations ($n = 20000$). We also present the corresponding space densities (rights). The segregation sequences needs a special construction. The order of reaction must be chosen in despondence of the spatial positions of all particles. It is important to note also that the final distribution of the particles have similar behavior. The condensing process breaks down by reaching a minimal total energy of the collection of the particles (fixed with a tolerance). The densities have a stochastic diffusion behavior, such stochastic diffusion problem was studied in the PDE case by Manouzi and al. [24].

For $\varepsilon = 0.1, \theta = 0.001$ and NP=400, we chose a random (uniform) initial distribution of masses. The figures 2.10 present four iterations of a segregation, we show the iterations 400, 6000 and 20000. It looks like a stochastic diffusion of mass. (the density is computed only in $[0, 1]^2$. The figures2.11 show the special moves of particles, the number of the support of m^i in each iteration is constant. It looks like a diffusion of density.

(b) Simulation of segregation of a measure with random distribution of mass:

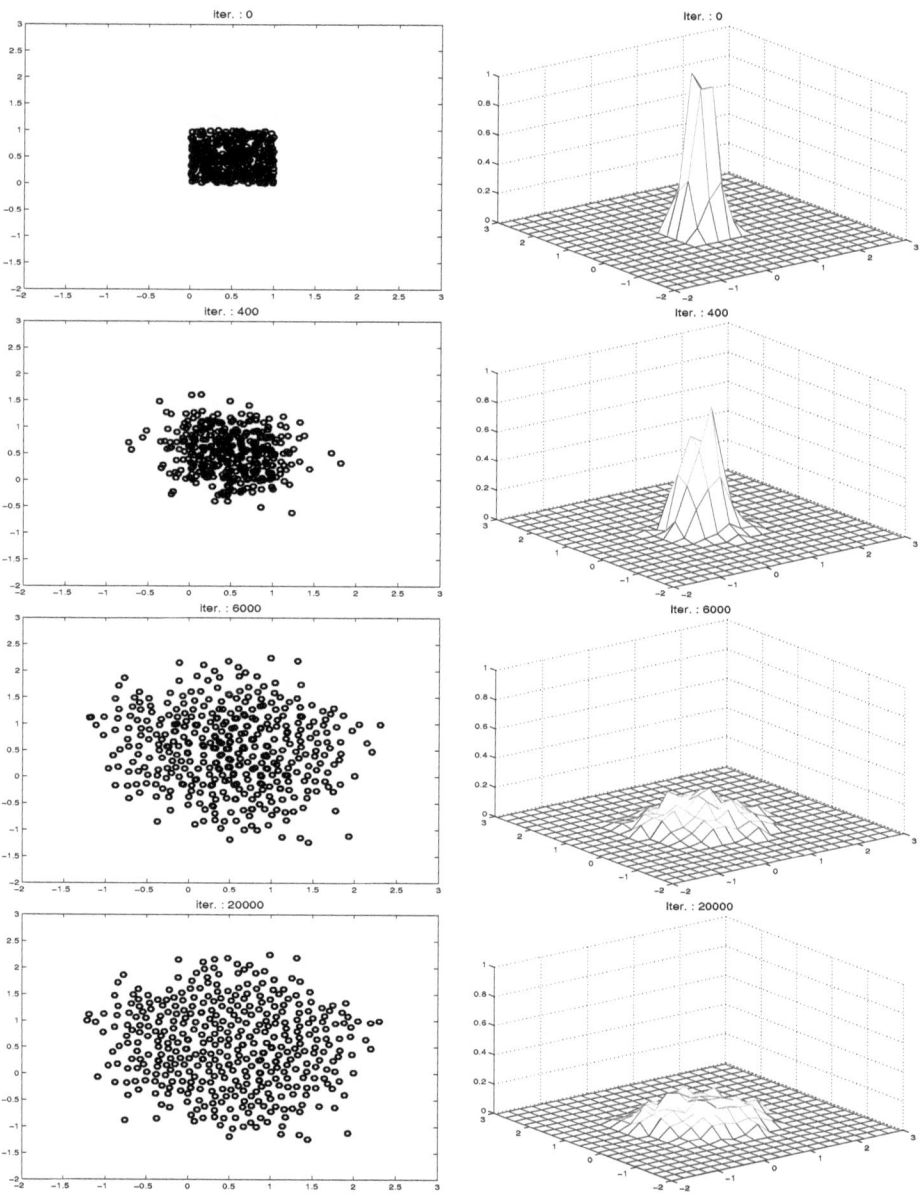

Figure 2.10: *Density (right) and spatial position (left) of segregation. (sim. (a))*.

2.3. Numerical Simulations

(b) Simulation of segregation of a measure with deterministic mass:

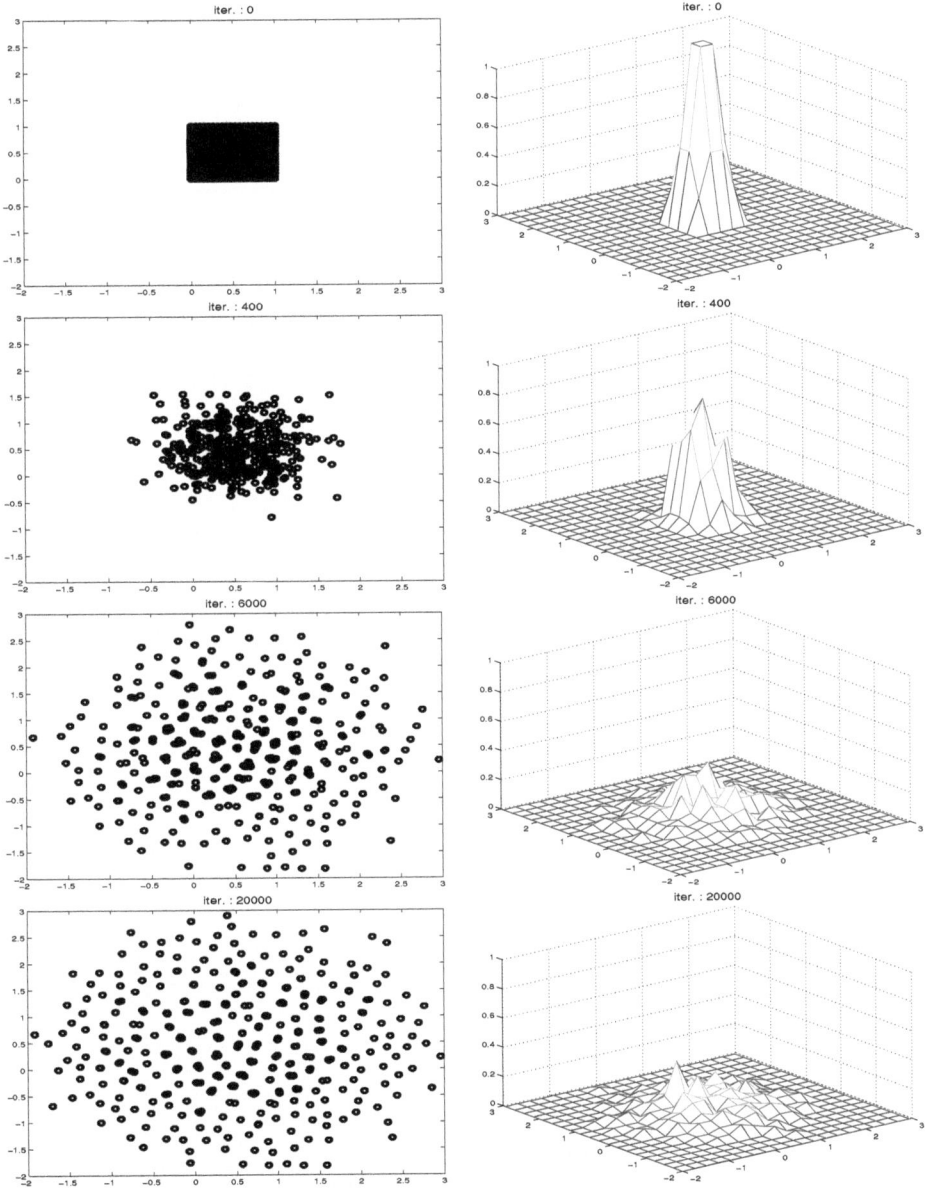

Figure 2.11: *Density (right) and spatial position (left) of segregation. (sim. (b)).*

2.3.5 Simultaneously condensing on the unit circle

For $NP = 100$ random masses chosen with a uniform random distribution $U(0,1)$, we simulate simultaneously condensing sequences on the unit circle. The first one is the average model (HK model) and the second one is the energy model. For $\varepsilon = \frac{\pi}{8}$, $\varepsilon = \frac{\pi}{16}$ and $\theta = 10^{-4}$ (only for the energy model), we run our code for a measure $m := \sum_{x \in S(m) \subset S^1} m(x)\delta_x$, where $\sharp S(m) = 100$. The figures 2.12 and 2.13 present a time iteration plot of simultaneously condensing sequences on the unit circle (as one dimensional manifold). The following table summarizes the data for this simulation:

Parameter/Sim.	Energy model	HK model
NP	100	100
ε	$\frac{\pi}{8}$ and $\frac{\pi}{16}$	$\frac{\pi}{8}$ and $\frac{\pi}{16}$
θ	10^{-4}	–
Initial state	100 masses $U(0,1)$	100 masses $U(0,1)$
Final state	isolated masses	isolated masses

Table 2.5: Results of simultaneously condensing sequences in S^1.

Remark on the results 2.3.5. Generally, we have remarked that the limit state (convergence) using the average model is faster than the energy model. The main characteristic of the average model is that the particles maintain the same position in different groups with distance greater than or equal to epsilon, but in the energy model the particles does not maintain the same positions: A jump moves is possible (see the right plot of 2.13, which don't guaranties the the convergence). According to this simulation, consider the two particles, which are simultaneously moving and changing they positions in a periodic scenario. All this open very nice question, namely under which conditions converges a simultaneously condensing sequence.

2.3. Numerical Simulations

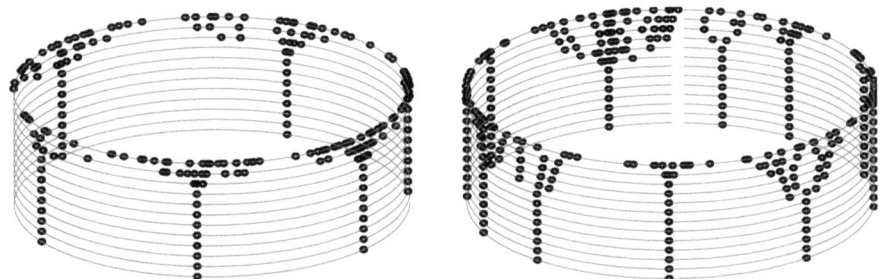

Figure 2.12: *Simultaneously condensing on the unit circle for the average model.*

The figures present the condensing process of particles on the unit circle using the average model (HK model). We show eight time iterations of simultaneously condensing of particles. The convergence is achieved after four iterations for $\varepsilon = \frac{\pi}{8}$ and after five iterations for $\varepsilon = \frac{\pi}{16}$ and we also show that the limit state is collection of point masses.

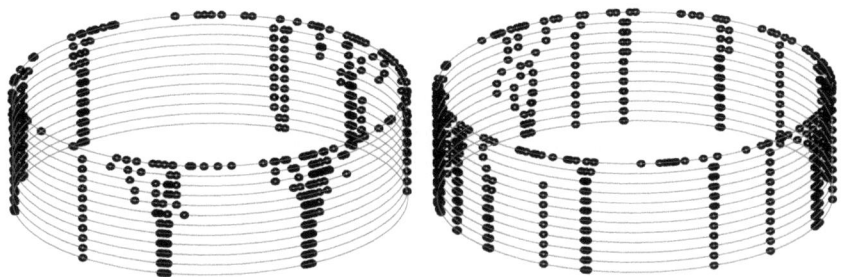

Figure 2.13: *Simultaneously condensing on the unit circle for the energy model.*

The figures present the condensing process of particles on the unit circle using the energy model (condensing model). We show eight time iterations of simultaneously condensing of particles for $\varepsilon = \frac{\pi}{8}$ and $\varepsilon = \frac{\pi}{16}$ respectively.

Concluding remarks

In the present work we have proposed a new model for condensing sequences, with special interest on the condensing process of particles. In one hand, we have shown how a collection of particles with a local control rule, forms an isolated distribution of masses with zero global energy. In the other hand, we have analyzed the model in different metric spaces. The results of the condensing process in a finite metric space are also simulated with respect to random metrics, which is a suggestion of Prof. Dr. Sieveking to our best knowledge, this is the first time that random metrics are implemented and experimented for a class of metric spaces, namely finite metric spaces. Our main concern is to extent the idea of performing simulations with respect to random metrics for many numerical phenomena. The present study can only be considered as and example for explaining the concept of consensus and emergence phenomena. It should be stressed that the stochastic behavior of our simulations is due to the random choice of positions minimizing the local energy.

From a computational viewpoint, the implementation of the condensing algorithm in a discrete metric space is very demanding in terms of the number of operations. This complexity is mainly because of the specific definition of the metric spaces. For instance, one has to store many information to be used for choosing positions with a minimal energy. In our implementation, we have used a serial `C++` code for computations and a `Matlab` routine for graphical visualizations.

For the random generators used in the computations, we used those proposed in my diploma thesis carried out at the technical university of Darmstadt [25].

The proposed condensing model can be generalized to metric space with complex structure (complex surface) without major conceptual modifications. As an application, we have applied the proposed condensing model in finite metric spaces. The idea of discrete metrics can also be extended to approximate manifold spaces with a complex structure. Furthermore, using a stochastic energy-based model. In this study, to explain the condensing process, we have limited our work to simple academic and standard examples.

Although we have restricted our model and numerical computations to the time discrete displacements and discrete measures, the more important extension of our study concerns the use of continuous measures with nonnegative density and also the use of the time continuous displacements. As an example, one can consider the dynamical system in which the position occupied by a member of a group is determined by a continuous trajectory. One of our future works is therefore, to implement this method for solving the gradient problem and also for a continuous measure.

As a final remark, we have observed that using the model proposed by Hegselmann and Krause in [15] with respect to random metrics, the obtained energy will be periodic up to a deterministic range. The question is, under which conditions one can insure the decrease the total energy of this model? and how can one use the singular condensing to prove the convergence of the HK model? Furthermore, for which condensing mapping sequence f_i the energy $E(f_i(m))$ will be decease? More interesting problem is the construction and analysis and the convergence of the extended condensing model in a measure with density function?. Finally, how can one prove this convergence by modeling the trajectories of the particles as a gradient system in a continuous metric space? In conclusion, using techniques developed in the first chapter of the present work, one is able to simulate random metrics in a finite metric space. Therefore, generalization of these techniques for a random metric in arbitrary metric spaces (*i.e.* continuous metric spaces) would be of great interest.

Bibliography

[1] Babak P., MagnússonK. G., and S. Th. Sigurdsson.**Dynamics of group formation in collective motion of organisms**. Math. Med. Biol.,21:269–292, 2004.

[2] Bandelt H. -J. and Dress A., **A canonical split decomposition theory for metrics on a finite set** , AiM 92/47–105, 1992.

[3] Barbaro A., Birnir B., and Taylor K., **A comparison of simulations of discrete and ode models for the motion of pelagic fish**. UCSB Preprint, 2006.

[4] Barbaro A., Birnir B., and Taylor K.,**Discrete and Continuous Models of the Dynamics of Pelagic Fish: Orientation-Induced Swarming and Applications to the Capelin**. 2007.

[5] Birnir B., Bonilla L., and Soler J., **Complex fish schools**. UCSB Preprint, 2006.

[6] Birnir B., **An ODE Model of the Motion of Pelagic Fish**. Department of Mathematics University of California, Santa Barbara 2007.

[7] Birnir B., Einarsson B., and Sigurdsson S.,**Simulations and stability of complex fish schools**. RH Preprint, 2006.

[8] Breder C. M., **Equations descriptive of fish schools and other animal aggregations**. Ecology,vol. 35, pp. 361–370, 1954.

[9] Cucker F., Smale S., **Emergent Behavior in Flocks**. Automatic Control, IEEE Transactions on, 2007

[10] Deza M. and Laurent: **Geometry of cuts and metrics, Algorithm and Combinatorics**, 15 Springer-Verlag, Berlin, 1997.

[11] Dutour M., **Cut and Metrics Cones, Classification of facets and extrems rays for** $n \geq 8$, posted at http://www.liga.ens.fr/dutour/CUT-MET/CutMetricsCones.html.

[12] Dress M., **Trees, tight extensions of metric spaces, and the cohomological dimension of certain groups**: a note on combinatorial properties of metric spaces, Advances in Mathematics 53 no. 3, 321–402, 1984.

[13] Edelstein-Keshet L., **Mathematical models of swarming and social aggregation**. Dept of Mathematics, UBC Vancouver, BC Canada, 2001.

[14] Gazi V., B. Fidan, Y. S. Hanay And M. I. Köksal, **Aggregation, Foraging, and Formation Control of Swarms with Non-Holonomic Agents Using Potential Functions and Sliding Mode Techniques**. Turk J Elec Engin, VOL.15, NO.2, 2007.

[15] Hegselmann R., Krause U., **Opinion Dynamics and Bounded Confidence, Models, Analysis and Simulation**. Journal of Artificial Societies and Social Simulation, 2001.

[16] Hegselmann R and Krause U, **Opinion Dynamics Driven by Various Ways of Averaging**. Computational Economics 25. pp. 381–405, 2005.

[17] Hegselmann R and Krause U, **Truth and Cognitive Division of Labour: First Steps Towards a Computer Aided Social Epistemology**. Journal of Artificial Societies and Social Simulation vol. 9, no. 3, 2006.

[18] Krause U, **Soziale Dynamiken mit vielen Interakteuren. Eine Problemskizze. In Krause U and Stöckler M (Eds.) Modellierung und Simulation von Dynamiken mit vielen interagierenden Akteuren**. Universität Bremen, 1997.

[19] Lorenz J., **Consensus Strikes Back in the Hegselmann-Krause Model of Continuous Opinion Dynamics Under Bounded Confidence**. Journal of Artificial Societies and Social Simulation vol.9, no. 1, 2006.

[20] Shen J., **Cucker-Smale Flocking under Hierarchical Leadership**. Populations and Evolution(q-bio.PE); Quantitative Methods (q-bio.QM), 2006

[21] Sieveking M., **Condensing**. In Positve Systems (Proceedings of the second Multidisciplinary International Symposium on Positive Systems: Theory and Applications (POSTA 06) Grenoble, France, Aug. 3031, Sept. 1, 2006), volume 341 of Lecture Notes in Control and Information Sciences, pages 247–254, 2006.

[22] Herbert G. Tanner, Ali Jadbabaie2, and George J. Pappas, **Flocking in Teams of Nonholonomic Agents**. Lecture Notes in Control and Information Sciences, Springer Berlin / Heidelberg,229–239, 2005.

[23] M. Zahri, **Energy based Model for self organizing groups**, Proceeding, 1st Hispano-Moroccan Days on Applied Mathematics and Statistics, Vol 1, 411–416, 2008.

[24] Zahri M., Manouzi H., Seaid M., **Wick-Stochastic Finite Element Solution of Reaction-Diffusion Problems**, J. Comp. Applied Math. Vol 203, pp:516–532, 2007.

[25] Zahri M., **Analyse und Implementation des mehrdimensionalen Milstein-Verfahrens zur Simulation von SDE**, TU Darmstadt 2003.

Die VDM Verlagsservicegesellschaft sucht für wissenschaftliche Verlage abgeschlossene und herausragende

Dissertationen, Habilitationen, Diplomarbeiten, Master Theses, Magisterarbeiten usw.

für die kostenlose Publikation als Fachbuch.

Sie verfügen über eine Arbeit, die hohen inhaltlichen und formalen Ansprüchen genügt, und haben Interesse an einer honorarvergüteten Publikation?

Dann senden Sie bitte erste Informationen über sich und Ihre Arbeit per Email an *info@vdm-vsg.de*.

Sie erhalten kurzfristig unser Feedback!

VDM Verlagsservicegesellschaft mbH
Dudweiler Landstr. 99　　　　　　Telefon +49 681 3720 174
D - 66123 Saarbrücken　　　　　　Fax　　　+49 681 3720 1749
www.vdm-vsg.de

Die VDM Verlagsservicegesellschaft mbH vertritt

Printed by Books on Demand GmbH, Norderstedt / Germany